ON VIOLENCE

Hannah Arendt

ON VIOLENCE

A Harvest Book
Harcourt, Inc.
Orlando Austin New York San Diego London

For information about permission to reproduce selections from
this book, write to trade. permissions@hmhco.com or to Permissions,
Houghton Mifflin Harcourt Publishing Company, 3 Park Avenue,
19th Floor, New York, New York 10016.

www.hmhco.com

Library of Congress Catalog Card Number: 74-95867
ISBN 978-0-15-669500-8 (Harvest: pbk.)

Printed in the United States of America

DOC 50 49 48 47 46 45 44 43 42

4500746238

For Mary
in Friendship

I

THESE REFLECTIONS were provoked by the events and debates of the last few years as seen against the background of the twentieth century, which has become indeed, as Lenin predicted, a century of wars and revolutions, hence a century of that violence which is currently believed to be their common denominator. There is, however, another factor in the present situation which, though predicted by nobody, is of at least equal importance. The technical development of the implements of violence has now reached the point where no political goal could conceivably correspond to their destructive potential or justify their actual use in armed conflict. Hence, warfare—from time immemorial the final merciless arbiter in international disputes—has lost much of its effectiveness and nearly all its glamour. The "apocalyptic" chess game between the superpowers, that is, between those that move on the highest plane of our civilization, is being played according to the rule "if either 'wins' it is the end of both"; [1] it is a game that bears no resemblance to whatever war games preceded it. Its "rational" goal is deterrence, not victory, and the arms

[1] Harvey Wheeler, "The Strategic Calculators," in Nigel Calder, *Unless Peace Comes*, New York, 1968, p. 109.

race, no longer a preparation for war, can now be justified only on the grounds that more and more deterrence is the best guarantee of peace. To the question how shall we ever be able to extricate ourselves from the obvious insanity of this position, there is no answer.

Since violence—as distinct from power, force, or strength —always needs *implements* (as Engels pointed out long ago),[2] the revolution of technology, a revolution in toolmaking, was especially marked in warfare. The very substance of violent action is ruled by the means-end category, whose chief characteristic, if applied to human affairs, has always been that the end is in danger of being overwhelmed by the means which it justifies and which are needed to reach it. Since the end of human action, as distinct from the end products of fabrication, can never be reliably predicted, the means used to achieve political goals are more often than not of greater relevance to the future world than the intended goals.

Moreover, while the results of men's actions are beyond the actors' control, violence harbors within itself an additional element of arbitrariness; nowhere does Fortuna, good or ill luck, play a more fateful role in human affairs than on the battlefield, and this intrusion of the utterly unexpected does not disappear when people call it a "random event" and find it scientifically suspect; nor can it be eliminated by simulations, scenarios, game theories, and the like. There is no certainty in these matters, not even an ultimate certainty of mutual destruction under certain calculated circumstances. The very fact that those engaged in the perfection of the means of destruction have finally reached a level of technical development where their aim, namely, warfare, is on the point of dis-

[2] *Herrn Eugen Dührings Umwälzung der Wissenschaft* (1878), Part II, ch. 3.

4

appearing altogether by virtue of the means at its disposal [3] is like an ironical reminder of this all-pervading unpredictability, which we encounter the moment we approach the realm of violence. The chief reason warfare is still with us is neither a secret death wish of the human species, nor an irrepressible instinct of aggression, nor, finally and more plausibly, the serious economic and social dangers inherent in disarmament,[4] but the simple fact that no substitute for this final arbiter in international affairs has yet appeared on the political scene. Was not Hobbes right when he said: "Covenants, without the sword, are but words"?

Nor is a substitute likely to appear so long as national independence, namely, freedom from foreign rule, and the sovereignty of the state, namely, the claim to unchecked and unlimited power in foreign affairs, are identified. (The United States of America is among the few countries where a proper separation of freedom and sovereignty is at least theoretically possible insofar as the very

[3] As General André Beaufre, in "Battlefields of the 1980s," points out: Only "in those parts of the world not covered by nuclear deterrence" is war still possible, and even this "conventional warfare," despite its horrors, is actually already limited by the ever-present threat of escalation into nuclear war. (In Calder, *op. cit.*, p. 3.)

[4] *Report from Iron Mountain*, New York, 1967, the satire on the Rand Corporation's and other think tanks' way of thinking, is probably closer to reality, with its "timid glance over the brink of peace," than most "serious" studies. Its chief argument, that war is so essential to the functioning of our society that we dare not abolish it unless we discover even more murderous ways of dealing with our problems, will shock only those who have forgotten to what an extent the unemployment crisis of the Great Depression was solved only through the outbreak of World War II, or those who conveniently neglect or argue away the extent of present latent unemployment behind various forms of featherbedding.

5

foundations of the American republic would not be threatened by it. Foreign treaties, according to the Constitution, are part and parcel of the law of the land, and—as Justice James Wilson remarked in 1793—"to the Constitution of the United States the term sovereignty is totally unknown." But the times of such clearheaded and proud separation from the traditional language and conceptual political frame of the European nation-state are long past; the heritage of the American Revolution is forgotten, and the American government, for better and for worse, has entered into the heritage of Europe as though it were its patrimony—unaware, alas, of the fact that Europe's declining power was preceded and accompanied by political bankruptcy, the bankruptcy of the nation-state and its concept of sovereignty.) That war is still the *ultima ratio*, the old continuation of politics by means of violence, in the foreign affairs of the underdeveloped countries is no argument against its obsoleteness, and the fact that only small countries without nuclear and biological weapons can still afford it is no consolation. It is a secret from nobody that the famous random event is most likely to arise from those parts of the world where the old adage "There is no alternative to victory" retains a high degree of plausibility.

Under these circumstances, there are, indeed, few things that are more frightening than the steadily increasing prestige of scientifically minded brain trusters in the councils of government during the last decades. The trouble is not that they are cold-blooded enough to "think the unthinkable," but that they do not *think*. Instead of indulging in such an old-fashioned, uncomputerizable activity, they reckon with the consequences of certain hypothetically assumed constellations without, however, being able to test their hypotheses against actual occurrences. The logical flaw in these hypothetical constructions of future events is always the same: what first appears as a hypothesis—with or without its implied alternatives, according to the level of sophistication—turns immediately, usually after a few paragraphs, into a "fact," which then gives birth to a whole string of similar non-facts, with the result that the purely speculative character of the whole enterprise is forgotten. Needless to say, this is not science but pseudo-science, "the desperate attempt of the social and behavioral sciences," in the words of Noam Chomsky, "to imitate the surface features of sciences that really have significant intellectual content." And the most obvious and "most profound objection to this kind of strategic theory is not its limited usefulness but its danger, for it can lead us to believe we have an understanding of events and control over their flow which we do not have," as Richard N. Goodwin recently pointed out in a review article that had the rare virtue of detecting the "unconscious humor" characteristic of many of these pompous pseudo-scientific theories.[5]

Events, by definition, are occurrences that interrupt routine processes and routine procedures; only in a world in which nothing of importance ever happens could the futurologists' dream come true. Predictions of the future are never anything but projections of present automatic processes and procedures, that is, of occurrences that are likely to come to pass if men do not act and if nothing unexpected happens; every action, for better or worse, and every accident necessarily destroys the whole pattern in whose frame the prediction moves and where it finds its evidence. (Proudhon's passing remark, "The fecundity of the unexpected far exceeds the statesman's prudence," is

[5] Noam Chomsky in *American Power and the New Mandarins*, New York, 1969; Richard N. Goodwin's review of Thomas C. Schelling's *Arms and Influence*, Yale, 1966, in *The New Yorker*, February 17, 1968.

fortunately still true. It exceeds even more obviously the expert's calculations.) To call such unexpected, unpredicted, and unpredictable happenings "random events" or "the last gasps of the past," condemning them to irrelevance or the famous "dustbin of history," is the oldest trick in the trade; the trick, no doubt, helps in clearing up the theory, but at the price of removing it further and further from reality. The danger is that these theories are not only plausible, because they take their evidence from actually discernible present trends, but that, because of their inner consistency, they have a hypnotic effect; they put to sleep our common sense, which is nothing else but our mental organ for perceiving, understanding, and dealing with reality and factuality.

No one engaged in thought about history and politics can remain unaware of the enormous role violence has always played in human affairs, and it is at first glance rather surprising that violence has been singled out so seldom for special consideration.[6] (In the last edition of the Encyclopedia of the Social Sciences "violence" does not even rate an entry.) This shows to what an extent violence and its arbitrariness were taken for granted and therefore neglected; no one questions or examines what is obvious to all. Those who saw nothing but violence in human affairs, convinced that they were "always haphazard, not serious, not precise" (Renan) or that God was forever with the bigger battalions, had nothing more to say about either violence or history. Anybody looking for some kind of sense in the records of the past was almost bound to see violence as a marginal phenomenon. Whether it is Clausewitz calling war "the continuation of politics by other

---

[6] There exists, of course, a large literature on war and warfare, but it deals with the implements of violence, not with violence as such.

means," or Engels defining violence as the accelerator of economic development,[7] the emphasis is on political or economic continuity, on the continuity of a process that remains determined by what preceded violent action. Hence, students of international relations have held until recently that "it was a maxim that a military resolution in discord with the deeper cultural sources of national power could not be stable," or that, in Engels' words, "wherever the power structure of a country contradicts its economic development" it is political power with its means of violence that will suffer defeat.[8]

Today all these old verities about the relation between war and politics or about violence and power have become inapplicable. The Second World War was not followed by peace but by a cold war and the establishment of the military-industrial-labor complex. To speak of "the priority of war-making potential as the principal structuring force in society," to maintain that "economic systems, political philosophies, and corpora juris serve and extend the war system, not vice versa," to conclude that "war itself is the basic social system, within which other secondary modes of social organization conflict or conspire"—all this sounds much more plausible than Engels' or Clausewitz's nineteenth-century formulas. Even more conclusive than this simple reversal proposed by the anonymous author of the *Report from Iron Mountain*—instead of war being "an extension of diplomacy (or of politics, or of the pursuit of economic objectives)," peace is the continuation of war by other means—is the actual development in the techniques of warfare. In the words of the Russian physicist Sakharov, "A thermonuclear war cannot be considered a continuation of politics by other means (according to the

[7] See Engels, *op. cit.*, Part II, ch. 4.

[8] Wheeler, *op. cit.*, p. 107; Engels, *ibidem*.

9

formula of Clausewitz). It would be a means of universal suicide." [9]

Moreover, we know that "a few weapons could wipe out all other sources of national power in a few moments," [10] that biological weapons have been devised which would enable "small groups of individuals . . . to upset the strategic balance" and would be cheap enough to be produced by "nations unable to develop nuclear striking forces," [11] that "within a very few years" robot soldiers will have made "human soldiers completely obsolete," [12] and that, finally, in conventional warfare the poor countries are much less vulnerable than the great powers precisely because they are "underdeveloped," and because technical superiority can "be much more of a liability than an asset" in guerrilla wars.[13] What all these uncomfortable novelties add up to is a complete reversal in the relationship between power and violence, foreshadowing another reversal in the future relationship between small and great powers. The amount of violence at the disposal of any given country may soon not be a reliable indication of the country's strength or a reliable guarantee against destruction by a substantially smaller and weaker power. And this bears an ominous similarity to one of political science's oldest insights, namely that power cannot be measured in terms of wealth, that an abundance of wealth may erode power, that riches are particularly dangerous to the power

[9] Andrei D. Sakharov, *Progress, Coexistence, and Intellectual Freedom,* New York, 1968, p. 36.

[10] Wheeler, *ibidem.*

[11] Nigel Calder, "The New Weapons," in *op. cit.,* p. 239.

[12] M. W. Thring, "Robots on the March," in Calder, *op. cit.,* p. 169.

[13] Vladimir Dedijer, "The Poor Man's Power," in Calder, *op. cit.,* p. 29.

and well-being of republics—an insight that does not lose in validity because it has been forgotten, especially at a time when its truth has acquired a new dimension of validity by becoming applicable to the arsenal of violence as well.

The more dubious and uncertain an instrument violence has become in international relations, the more it has gained in reputation and appeal in domestic affairs, specifically in the matter of revolution. The strong Marxist rhetoric of the New Left coincides with the steady growth of the entirely non-Marxian conviction, proclaimed by Mao Tse-tung, that "Power grows out of the barrel of a gun." To be sure, Marx was aware of the role of violence in history, but this role was to him secondary; not violence but the contradictions inherent in the old society brought about its end. The emergence of a new society was preceded, but not caused, by violent outbreaks, which he likened to the labor pangs that precede, but of course do not cause, the event of organic birth. In the same vein he regarded the state as an instrument of violence in the command of the ruling class; but the actual power of the ruling class did not consist of or rely on violence. It was defined by the role the ruling class played in society, or, more exactly, by its role in the process of production. It has often been noticed, and sometimes deplored, that the revolutionary Left under the influence of Marx's teachings ruled out the use of violent means; the "dictatorship of the proletariat"—openly repressive in Marx's writings—came after the revolution and was meant, like the Roman dictatorship, to last a strictly limited period. Political assassination, except for a few acts of individual terror perpetrated by small groups of anarchists, was mostly the prerogative of the Right, while organized armed uprisings remained the specialty of the military. The Left remained convinced "that all conspiracies are not only useless but harmful.

11

They [knew] only too well that revolutions are not made intentionally and arbitrarily, but that they were always and everywhere the necessary result of circumstances entirely independent of the will and guidance of particular parties and whole classes." [14]

On the level of theory there were a few exceptions. Georges Sorel, who at the beginning of the century tried to combine Marxism with Bergson's philosophy of life—the result, though on a much lower level of sophistication, is oddly similar to Sartre's current amalgamation of existentialism and Marxism—thought of class struggle in military terms; yet he ended by proposing nothing more violent than the famous myth of the general strike, a form of action which we today would think of as belonging rather to the arsenal of nonviolent politics. Fifty years ago even this modest proposal earned him the reputation of being a fascist, notwithstanding his enthusiastic approval of Lenin and the Russian Revolution. Sartre, who in his preface to Fanon's *The Wretched of the Earth* goes much farther in his glorification of violence than Sorel in his famous *Reflections on Violence*—farther than Fanon himself, whose argument he wishes to bring to its conclusion—still mentions "Sorel's fascist utterances." This shows to what extent Sartre is unaware of his basic disagreement with Marx on the question of violence, especially when he states that "irrepressible violence . . . is man recreating himself," that it is through "mad fury" that "the wretched of the earth" can "become men." These notions are all the more remarkable because the idea of man creating himself is strictly in the tradition of Hegelian and Marxian thinking; it is the very basis of all leftist humanism. But according to Hegel man "produces" himself through

[14] I owe this early remark of Engels, in a manuscript of 1847, to Jacob Barion, *Hegel und die marxistische Staatslehre*, Bonn, 1963.

thought,[15] whereas for Marx, who turned Hegel's "idealism" upside down, it was labor, the human form of metabolism with nature, that fulfilled this function. And though one may argue that all notions of man creating himself have in common a rebellion against the very factuality of the human condition—nothing is more obvious than that man, whether as member of the species or as an individual, does *not* owe his existence to himself—and that therefore what Sartre, Marx, and Hegel have in common is more relevant than the particular activities through which this non-fact should presumably have come about, still it cannot be denied that a gulf separates the essentially peaceful activities of thinking and laboring from all deeds of violence. "To shoot down a European is to kill two birds with one stone . . . there remain a dead man and a free man," says Sartre in his preface. This is a sentence Marx could never have written.[16]

I quoted Sartre in order to show that this new shift toward violence in the thinking of revolutionaries can remain unnoticed even by one of their most representative and articulate spokesmen,[17] and it is all the more noteworthy for evidently not being an abstract notion in the history of ideas. (If one turns the "idealistic" *concept* of thought upside down, one might arrive at the "materialistic" *concept* of labor; one will never arrive at the notion of violence.) No doubt all this has a logic of its own, but it is one springing from experience, and this experience was utterly unknown to any generation before.

The pathos and the *élan* of the New Left, their credi-

[15] It is quite suggestive that Hegel speaks in this context of "*Sichselbstproduzieren.*" See *Vorlesungen über die Geschichte der Philosophie,* ed. Hoffmeister, p. 114, Leipzig, 1938.

[16] See appendix I, p. 89.

[17] See appendix II, p. 89.

bility, as it were, are closely connected with the weird suicidal development of modern weapons; this is the first generation to grow up under the shadow of the atom bomb. They inherited from their parents' generation the experience of a massive intrusion of criminal violence into politics: they learned in high school and in college about concentration and extermination camps, about genocide and torture,[18] about the wholesale slaughter of civilians in war without which modern military operations are no longer possible even if restricted to "conventional" weapons. Their first reaction was a revulsion against every form of violence, an almost matter-of-course espousal of a politics of nonviolence. The very great successes of this movement, especially in the field of civil rights, were followed by the resistance movement against the war in Vietnam, which has remained an important factor in determining the climate of opinion in this country. But it is no secret that things have changed since then, that the adherents of nonviolence are on the defensive, and it would be futile to say that only the "extremists" are yielding to a glorification of violence and have discovered—like Fanon's Algerian peasants—that "only violence pays." [19]

[18] Noam Chomsky rightly notices among the motives for open rebellion the refusal "to take one's place alongside the 'good German' we have all learned to despise." *Op. cit.*, p. 368.

[19] Frantz Fanon, *The Wretched of the Earth* (1961), Grove Press edition, 1968, p. 61. I am using this work because of its great influence on the present student generation. Fanon himself, however, is much more doubtful about violence than his admirers. It seems that only the book's first chapter, "Concerning Violence," has been widely read. Fanon knows of the "unmixed and total brutality [which], if not immediately combatted, invariably leads to the defeat of the movement within a few weeks" (p. 147).
For the recent escalation of violence in the student movement, see the instructive series "Gewalt" in the German news magazine

The new militants have been denounced as anarchists, nihilists, red fascists, Nazis, and, with considerably more justification, "Luddite machine smashers," [20] and the students have countered with the equally meaningless slogans of "police state" or "latent fascism of late capitalism," and, with considerably more justification, "consumer society." [21] Their behavior has been blamed on all kinds of social and psychological factors—on too much permissiveness in their upbringing in America and on an explosive reaction to too much authority in Germany and Japan, on the lack of freedom in Eastern Europe and too much freedom in the West, on the disastrous lack of jobs for sociology students in France and the superabundance of careers in nearly all fields in the United States—all of which appear locally plausible enough but are clearly contradicted by the fact that the student rebellion is a global phenomenon. A social common denominator of the movement seems out of the question, but it is true that psychologically this generation seems everywhere char-

---

Der Spiegel (February 10, 1969 ff.), and the series "Mit dem Latein am Ende" (Nos. 26 and 27, 1969).

[20] See appendix III, p. 91.

[21] The last of these epithets would make sense if it were meant descriptively. Behind it, however, stands the illusion of Marx's society of free producers, the liberation of the productive forces of society, which in fact has been accomplished not by the revolution but by science and technology. This liberation, furthermore, is not accelerated, but seriously retarded, in all countries that have gone through a revolution. In other words, behind their denunciation of consumption stands the idealization of production, and with it the old idolization of productivity and creativity. "The joy of destruction is a creative joy"—yes indeed, if one believes that "the joy of labor" is productive; destruction is about the only "labor" left that can be done by simple implements without the help of machines, although machines do the job, of course, much more efficiently.

acterized by sheer courage, an astounding will to action, and by a no less astounding confidence in the possibility of change.[22] But these qualities are not causes, and if one asks what has actually brought about this wholly unexpected development in universities all over the world, it seems absurd to ignore the most obvious and perhaps the most potent factor, for which, moreover, no precedent and no analogy exist—the simple fact that technological progress" is leading in so many instances straight into disaster;[23] that the sciences, taught and learned by this generation, seem not merely unable to undo the disastrous consequences of their own technology but have reached a stage in their development where "there's no damn thing you can do that can't be turned into war." [24] (To be sure, nothing is more important to the integrity of the universities—which, in Senator Fulbright's words, have betrayed a public trust when they became dependent on gov-

[22] This appetite for action is especially noticeable in small and relatively harmless enterprises. Students struck successfully against campus authorities who were paying employees in the cafeteria and in buildings and grounds less than the legal minimum. The decision of the Berkeley students to join the fight for transforming an empty university-owned lot into a "People's Park" should be counted among these enterprises, even though it provoked the worst reaction so far from the authorities. To judge from the Berkeley incident, it seems that precisely such "nonpolitical" actions unify the student body behind a radical vanguard. "A student referendum, which saw the heaviest turnout in the history of student voting, found 85 percent of the nearly 15,000 who voted favoring the use of the lot" as a people's park. See the excellent report by Sheldon Wolin and John Schaar, "Berkeley: The Battle of People's Park," *New York Review of Books,* June 19, 1969.

[23] See appendix IV, p. 92.

[24] Thus Jerome Lettvin, of M.I.T., in the New York *Times Magazine,* May 18, 1969.

ernment-sponsored research projects [25]—than a rigorously enforced divorce from war-oriented research and all connected enterprises; but it would be naïve to expect this to change the nature of modern science or hinder the war effort, naïve also to deny that the resulting limitation might well lead to a lowering of university standards.[26] The only thing this divorce is not likely to lead to is a general withdrawal of federal funds; for, as Jerome Lettvin, of M.I.T., recently pointed out, "The Government can't afford not to support us" [27]—just as the universities cannot afford not to accept federal funds; but this means no more than that they "must learn how to sterilize financial support" (Henry Steele Commager), a difficult but not impossible task in view of the enormous increase of the power of universities in modern societies.) In short, the seemingly irresistible proliferation of techniques and machines, far from only threatening certain classes with unemployment, menaces the existence of whole nations and conceivably of all mankind.

It is only natural that the new generation should live with greater awareness of the possibility of doomsday than those "over thirty," not because they are younger but because this was their first decisive experience in the world. (What are "problems" to us "are built into the flesh and blood of the young.") [28] If you ask a member of this generation two simple questions: "How do you want the world to be in fifty years?" and "What do you want your life to be like five years from now?" the answers are quite

[25] See appendix V, p. 93.

[26] The steady drift of basic research from the universities to the industrial laboratories is very significant and a case in point.

[27] *Loc. cit.*

[28] Stephen Spender, *The Year of the Young Rebels,* New York, 1969, p. 179.

17

often preceded by "Provided there is still a world," and "Provided I am still alive." In George Wald's words, "what we are up against is a generation that is by no means sure that it has a future." [29] For the future, as Spender puts it, is "like a time-bomb buried, but ticking away, in the present." To the often-heard question Who are they, this new generation? one is tempted to answer, Those who hear the ticking. And to the other question, Who are they who utterly deny them? the answer may well be, Those who do not know, or refuse to face, things as they really are.

The student rebellion is a global phenomenon, but its manifestations vary, of course, greatly from country to country, often from university to university. This is especially true of the practice of violence. Violence has remained mostly a matter of theory and rhetoric where the clash between generations did not coincide with a clash of tangible group interests. This was notably so in Germany, where the tenured faculty had a vested interest in overcrowded lectures and seminars. In America, the student movement has been seriously radicalized wherever police and police brutality intervened in essentially nonviolent demonstrations: occupations of administration buildings, sit-ins, et cetera. Serious violence entered the scene only with the appearance of the Black Power movement on the campuses. Negro students, the majority of them admitted without academic qualification, regarded and organized themselves as an interest group, the representatives of the black community. Their interest was to lower academic standards. They were more cautious than the white rebels, but it was clear from the beginning (even before the incidents at Cornell University and City College in New York) that violence with them was not a matter of theory and rhetoric. Moreover, while the student rebellion in

[29] George Wald in *The New Yorker*, March 22, 1969.

Western countries can nowhere count on popular support outside the universities and as a rule encounters open hostility the moment it uses violent means, there stands a large minority of the Negro community behind the verbal or actual violence of the black students.[30] Black violence can indeed be understood in analogy to the labor violence in America a generation ago; and although, as far as I know, only Staughton Lynd has drawn the analogy between labor riots and student rebellion explicitly,[31] it seems that the academic establishment, in its curious tendency to yield more to Negro demands, even if they are clearly silly and outrageous,[32] than to the disinterested and usually highly moral claims of the white rebels, also thinks in these terms and feels more comfortable when confronted with interests plus violence than when it is a matter of nonviolent "participatory democracy." The yielding of university authorities to black demands has often been explained by the "guilt feelings" of the white community; I think it is more likely that faculty as well as administrations and boards of trustees are half-consciously aware of the obvious truth of a conclusion of the official *Report on Violence in America:* "Force and violence are likely to be successful techniques of social control and persuasion when they have wide popular support."[33]

The new undeniable glorification of violence by the student movement has a curious peculiarity. While the rheto-

[30] See appendix VI, p. 94.

[31] See appendix VII, p. 95.

[32] See appendix VIII, p. 95.

[33] See the report of the *National Commission on the Causes and Prevention of Violence,* June, 1969, as quoted from the New York *Times,* June 6, 1969.

19

ric of the new militants is clearly inspired by Fanon, their theoretical arguments contain usually nothing but a hodgepodge of all kinds of Marxist leftovers. This is indeed quite baffling for anybody who has ever read Marx or Engels. Who could possibly call an ideology Marxist that has put its faith in "classless idlers," believes that "in the lumpenproletariat the rebellion will find its urban spearhead," and trusts that "gangsters will light the way for the people"? [34] Sartre with his great felicity with words has given expression to the new faith. "Violence," he now believes, on the strength of Fanon's book, "like Achilles' lance, can heal the wounds it has inflicted." If this were true, revenge would be the cure-all for most of our ills. This myth is more abstract, farther removed from reality, than Sorel's myth of a general strike ever was. It is on a par with Fanon's worst rhetorical excesses, such as, "hunger with dignity is preferable to bread eaten in slavery." No history and no theory is needed to refute this statement; the most superficial observer of the processes that go on in the human body knows its untruth. But had he said that bread eaten with dignity is preferable to cake eaten in slavery the rhetorical point would have been lost.

Reading these irresponsible grandiose statements—and those I quoted are fairly representative, except that Fanon still manages to stay closer to reality than most—and looking at them in the perspective of what we know about the history of rebellions and revolutions, one is tempted to deny their significance, to ascribe them to a passing mood, or to the ignorance and nobility of sentiment of people exposed to unprecedented events and developments without any means of handling them mentally, and who therefore curiously revive thoughts and emotions from which Marx had hoped to liberate the revolution once and for all.

[34] Fanon, *op. cit.*, pp. 130, 129, and 69, respectively.

Who has ever doubted that the violated dream of violence, that the oppressed "dream at least once a day of setting" themselves up in the oppressor's place, that the poor dream of the possessions of the rich, the persecuted of exchanging "the role of the quarry for that of the hunter," and the last of the kingdom where "the last shall be first, and the first last"? [35] The point, as Marx saw it, is that dreams never come true.[36] The rarity of slave rebellions and of uprisings among the disinherited and downtrodden is notorious; on the few occasions when they occurred it was precisely "mad fury" that turned dreams into nightmares for everybody. In no case, as far as I know, was the force of these "volcanic" outbursts, in Sartre's words, "equal to that of the pressure put on them." To identify the national liberation movements with such outbursts is to prophesy their doom—quite apart from the fact that the unlikely victory would not result in changing the world (or the system), but only its personnel. To think, finally, that there is such a thing as a "Unity of the Third World," to which one could address the new slogan in the era of decolonization "Natives of all underdeveloped countries unite!" (Sartre) is to repeat Marx's worst illusions on a greatly enlarged scale and with considerably less justification. The Third World is not a reality but an ideology.[37]

[35] Fanon, *op. cit.*, pp. 37 ff., 53.

[36] See appendix IX, p. 96.

[37] The students caught between the two superpowers and equally disillusioned by East and West, "inevitably pursue some third ideology, from Mao's China or Castro's Cuba." (Spender, *op. cit.*, p. 92.) Their calls for Mao, Castro, Che Guevara, and Ho Chi Minh are like pseudo-religious incantations for saviors from another world; they would also call for Tito if only Yugoslavia were farther away and less approachable. The case is different with the Black Power movement; its ideological commitment to the nonexistent

The question remains why so many of the new preachers of violence are unaware of their decisive disagreement with Karl Marx's teachings, or, to put it another way, why they cling with such stubborn tenacity to concepts and doctrines that have not only been refuted by factual developments but are clearly inconsistent with their own politics. The one positive political slogan the new movement has put forth, the claim for "participatory democracy" that has echoed around the globe and constitutes the most significant common denominator of the rebellions in the East and the West, derives from the best in the revolutionary tradition—the council system, the always defeated but only authentic outgrowth of every revolution since the eighteenth century. But no reference to this goal either in word or substance can be found in the teachings of Marx and Lenin, both of whom aimed on the contrary at a society in which the need for public action and participation in public affairs would have "withered away," [38]

"Unity of the Third World" is not sheer romantic nonsense. They have an obvious interest in a black-white dichotomy; this too is of course mere escapism—an escape into a dream world in which Negroes would constitute an overwhelming majority of the world's population.

[38] It seems as though a similar inconsistency could be charged to Marx and Lenin. Did not Marx glorify the Paris Commune of 1871, and did not Lenin want to give "all power to the *soviets*"? But for Marx the Commune was no more than a transitory organ of revolutionary action, "a lever for uprooting the economical foundations of . . . class rule," which Engels rightly identified with the likewise transitory "dictatorship of the Proletariat." (See *The Civil War in France*, in Karl Marx and F. Engels, *Selected Works*, London, 1950, Vol. I, pp. 474 and 440, respectively.) The case of Lenin is more complicated. Still, it was Lenin who emasculated the *soviets* and gave all power to the party.

together with the state. Because of a curious timidity in theoretical matters, contrasting oddly with its bold courage in practice, the slogan of the New Left has remained in a declamatory stage, to be invoked rather inarticulately against Western representative democracy (which is about to lose even its merely representative function to the huge party machines that "represent" not the party membership but its functionaries) and against the Eastern one-party bureaucracies, which rule out participation on principle.

Even more suprising in this odd loyalty to the past is the New Left's seeming unawareness of the extent to which the moral character of the rebellion—now a widely accepted fact [39]—clashes with its Marxian rhetoric. Nothing, indeed, about the movement is more striking than its disinterestedness; Peter Steinfels, in a remarkable article on the "French revolution 1968" in *Commonweal* (July 26, 1968), was quite right when he wrote: "Péguy might have been an appropriate patron for the cultural revolution, with his later scorn for the Sorbonne mandarinate [and] his formula, 'The social Revolution will be moral or it will

---

[39] "Their revolutionary idea," as Spender (*op. cit.*, p. 114) states, "is moral passion." Noam Chomsky (*op. cit.*, p. 368) quotes facts: "The fact is that most of the thousand draft cards and other documents turned in to the Justice Department on October 20 [1967] came from men who can escape military service but who insisted on sharing the fate of those who are less privileged." The same was true for any number of draft-resister demonstrations and sit-ins in the universities and colleges. The situation in other countries is similar. *Der Spiegel* describes, for instance, the frustrating and often humiliating conditions of the research assistants in Germany: *"Angesichts dieser Verhältnisse nimmt es geradezu wunder, dass die Assistenten nicht in der vordersten Front der Radikalen stehen."* (June 23, 1969, p. 58.) It is always the same story: Interest groups do not join the rebels.

not be.' " To be sure, every revolutionary movement has been led by the disinterested, who were motivated by compassion or by a passion for justice, and this, of course, is also true for Marx and Lenin. But Marx, as we know, had quite effectively tabooed these "emotions"—if today the establishment dismisses moral arguments as "emotionalism" it is much closer to Marxist ideology than the rebels—and had solved the problem of "disinterested" leaders with the notion of their being the vanguard of mankind, embodying the ultimate interest of human history.[40] Still, they too had first to espouse the nonspeculative, down-to-earth interests of the working class and to identify with it; this alone gave them a firm footing outside society. And this is precisely what the modern rebels have lacked from the beginning and have been unable to find despite a rather desperate search for allies outside the universities. The hostility of the workers in all countries is a matter of record,[41] and in the United States the complete collapse of any co-operation with the Black Power movement, whose students are more firmly rooted in their own community and therefore in a better bargaining position at the universities, was the bitterest disappointment for the white rebels. (Whether it was wise of the Black Power people to refuse to play the role of the proletariat for "disinterested" leaders of a different color is another question.) It is, not surprisingly, in Germany, the old home of the Youth movement, that a group of students now proposes

[40] See appendix X, p. 96.

[41] Czechoslovakia seems to be an exception. However, the reform movement for which the students fought in the first ranks was backed by the whole nation, without any class distinctions. Marxistically speaking, the students there, and probably in all Eastern countries, have too much, rather than too little, support from the community to fit the Marxian pattern.

to enlist "all organized youth groups" in their ranks.[42] The absurdity of this proposal is obvious.

I am not sure what the explanation of these inconsistencies will eventually turn out to be; but I suspect that the deeper reason for this loyalty to a typically nineteenth-century doctrine has something to do with the concept of Progress, with an unwillingness to part with a notion that used to unite Liberalism, Socialism, and Communism into the "Left" but has nowhere reached the level of plausibility and sophistication we find in the writings of Karl Marx. (Inconsistency has always been the Achilles' heel of liberal thought; it combined an unswerving loyalty to Progress with a no less strict refusal to glorify History in Marxian and Hegelian terms, which alone could justify and guarantee it.)

The notion that there is such a thing as progress of mankind as a whole was unknown prior to the seventeenth century, developed into a rather common opinion among the eighteenth-century *hommes de lettres*, and became an almost universally accepted dogma in the nineteenth. But the difference between the earlier notions and their final stage is decisive. The seventeenth century, in this respect best represented by Pascal and Fontenelle, thought of progress in terms of an accumulation of knowledge through the centuries, whereas for the eighteenth the word implied an "education of mankind" (Lessing's *Erziehung des Menschengeschlechts*) whose end would coincide with man's coming of age. Progress was not unlimited, and Marx's classless society seen as the realm of freedom that could be the end of history—often interpreted as a secularization —of Christian eschatology or Jewish messianism—actually still bears the hallmark of the Age of Enlightenment. Be-

---

[42] See the Spiegel-Interview with Christoph Ehmann in *Der Spiegel*, February 10, 1969.

ginning with the nineteenth century, however, all such limitations disappeared. Now, in the words of Proudhon, motion is *"le fait primitif"* and "the laws of movement alone are eternal." This movement has neither beginning nor end: *"Le mouvement est; voilà tout!"* As to man, all we can say is "we are born perfectible, but we shall never be perfect." [43] Marx's idea, borrowed from Hegel, that every old society harbors the seeds of its successors in the same way every living organism harbors the seeds of its offspring is indeed not only the most ingenious but also the only possible conceptual guarantee for the sempiternal continuity of progress in history; and since the motion of this progress is supposed to come about through the clashes of antagonistic forces, it is possible to interpret every "regress" as a necessary but temporary setback.

To be sure, a guarantee that in the final analysis rests on little more than a metaphor is not the most solid basis to erect a doctrine upon, but this, unhappily, Marxism shares with a great many other doctrines in philosophy. Its great advantage becomes clear as soon as one compares it with other concepts of history—such as "eternal recurrences," the rise and fall of empires, the haphazard sequence of essentially unconnected events—all of which can equally be documented and justified, but none of which will guarantee a continuum of linear time and continuous progress in history. And the only competitor in the field, the ancient notion of a Golden Age at the beginning, from which everything else is derived, implies the rather unpleasant certainty of continuous decline. Of course, there are a few melancholy side effects in the reassuring idea that we need only march into the future,

[43] P.-J. Proudhon, *Philosophie du Progrès* (1853), 1946, pp. 27-30, 49, and *De la Justice* (1858), 1930, I, p. 238, respectively. See also William H. Harbold, "Progressive Humanity: in the Philosophy of P.-J. Proudhon," *Review of Politics*, January, 1969.

which we cannot help doing anyhow, in order to find a better world. There is first of all the simple fact that the general future of mankind has nothing to offer to individual life, whose only certain future is death. And if one leaves this out of account and thinks only in generalities, there is the obvious argument against progress that, in the words of Herzen, "Human development is a form of chronological unfairness, since late-comers are able to profit by the labors of their predecessors without paying the same price," [44] or, in the words of Kant, "It will always remain bewildering . . . that the earlier generations seem to carry on their burdensome business only for the sake of the later . . . and that only the last should have the good fortune to dwell in the [completed] building." [45]

However, these disadvantages, which were only rarely noticed, are more than outweighed by an enormous advantage: progress not only explains the past without breaking up the time continuum but it can serve as a guide for acting into the future. This is what Marx discovered when he turned Hegel upside down: he changed the direction of the historian's glance; instead of looking toward the past, he now could confidently look into the future. Progress gives an answer to the troublesome question, And what shall we do now? The answer, on the lowest level, says: Let us develop what we have into something better, greater, et cetera. (The, at first glance, irrational faith of liberals in growth, so characteristic of all our present political and economic theories, depends on this notion.) On the more sophisticated level of the Left, it tells us to develop present contradictions into their inherent synthesis.

[44] Alexander Herzen is quoted here from Isaiah Berlin's "Introduction" to Franco Venturi, *Roots of Revolutions*, New York, 1966.

[45] "Idea for a Universal History with Cosmopolitan Intent," Third Principle, in *The Philosophy of Kant,* Modern Library edition.

In either case we are assured that nothing altogether new and totally unexpected can happen, nothing but the "necessary" results of what we already know.[46] How reassuring that, in Hegel's words, "nothing else will come out but what was already there." [47]

I do not need to add that all our experiences in this century, which has constantly confronted us with the totally unexpected, stand in flagrant contradiction to these notions and doctrines, whose very popularity seems to consist in offering a comfortable, speculative or pseudo-scientific refuge from reality. A student rebellion almost exclusively inspired by moral considerations certainly belongs among the totally unexpected events of this century. This generation, trained like its predecessors in hardly anything but the various brands of the my-share-of-the-pie social and political theories, has taught us a lesson about manipulation, or, rather, its limits, which we would do well not to forget. Men can be "manipulated" through physical coercion, torture, or starvation, and their opinions can be arbitrarily formed by deliberate, organized misinformation, but not through "hidden persuaders," television, advertising, or any other psychological means in a free society. Alas, refutation of theory through reality has always been at best a lengthy and precarious business. The manipulation addicts, those who fear it unduly no less than those who have set their hopes on it, hardly notice when the chickens come home to roost. (One of the nicest examples of theories exploding into absurdity happened during the recent "People's Park" trouble in Berkeley.

[46] For an excellent discussion of the obvious fallacies in this position, see Robert A. Nisbet, "The Year 2000 and All That," in *Commentary*, June, 1968, and the ill-tempered critical remarks in the September issue.

[47] Hegel, *op. cit.*, p. 100 ff.

When the police and the National Guard, with rifles, unsheathed bayonets, and helicoptered riot gas, attacked the unarmed students—few of them "had thrown anything more dangerous than epithets"—some Guardsmen fraternized openly with their "enemies" and one of them threw down his arms and shouted: "I can't stand this any more." What happened? In the enlightened age we live in, this could be explained only by insanity; "he was rushed to a psychiatric examination [and] diagnosed as suffering from 'suppressed aggressions.' ") [48]

Progress, to be sure, is a more serious and a more complex item offered at the superstition fair of our time.[49] The irrational nineteenth-century belief in *unlimited* progress has found universal acceptance chiefly because of the astounding development of the natural sciences, which, since the rise of the modern age, actually have been "universal" sciences and therefore could look forward to an unending task in exploring the immensity of the universe. That science, even though no longer limited by the finitude of the earth and its nature, should be subject to never-ending progress is by no means certain; that strictly scientific research in the humanities, the so-called *Geisteswissenschaften* that deal with the products of the human spirit, must come to an end by definition is obvious. The ceaseless, senseless demand for original scholarship in a number of fields, where only erudition is now possible, has

[48] The incident is reported without comment by Wolin and Schaar, *op. cit.* See also Peter Barnes's report " 'An Outcry': Thoughts on Being Tear Gassed," in *Newsweek,* June 2, 1969.

[49] Spender (*op. cit.,* p. 45) reports that the French students during the May incidents in Paris "refused categorically the ideology of 'output' [*rendement*], of 'progress' and such-called pseudo-forces." In America, this is not yet the case as far as progress is concerned. We are still surrounded by talk about "progressive" and "regressive" forces, "progressive" and "repressive tolerance," and the like.

led either to sheer irrelevancy, the famous knowing of more and more about less and less, or to the development of a pseudo-scholarship which actually destroys its object.[50] It is noteworthy that the rebellion of the young, to the extent that it is not exclusively morally or politically motivated, has been chiefly directed against the academic glorification of scholarship and science, both of which, though for different reasons, are gravely compromised in their eyes. And it is true that it is by no means impossible that we have reached in both cases a turning point, the point of destructive returns. Not only has the progress of science ceased to coincide with the progress of mankind (whatever that may mean), but it could even spell mankind's end, just as the further progress of scholarship may well end with the destruction of everything that made scholarship worth our while. Progress, in other words, can no longer serve as the standard by which to evaluate the disastrously rapid change-processes we have let loose.

Since we are concerned here primarily with violence, I must warn against a tempting misunderstanding. If we look on history in terms of a continuous chronological process, whose progress, moreover, is inevitable, violence in the shape of war and revolution may appear to constitute the only possible interruption. If this were true, if only the practice of violence would make it possible to interrupt automatic processes in the realm of human affairs, the preachers of violence would have won an important point. (Theoretically, as far as I know, the point was never made, but it seems to me incontestable that the disruptive student activities in the last few years are actually based on this conviction.) It is the function, how-

[50] For a splendid exemplification of these not merely superfluous but pernicious enterprises, see Edmund Wilson, *The Fruits of the MLA*, New York, 1968.

ever, of all action, as distinguished from mere behavior, to interrupt what otherwise would have proceeded automatically and therefore predictably.

II

I T I S against the background of these experiences that I propose to raise the question of violence in the political realm. This is not easy; what Sorel remarked sixty years ago, "The problems of violence still remain very obscure," [51] is as true today as it was then. I mentioned the general reluctance to deal with violence as a phenomenon in its own right, and I must now qualify this statement. If we turn to discussions of the phenomenon of power, we soon find that there exists a consensus among political theorists from Left to Right to the effect that violence is nothing more than the most flagrant manifestation of power. "All politics is a struggle for power; the ultimate kind of power is violence," said C. Wright Mills, echoing, as it were, Max Weber's definition of the state as "the rule of men over men based on the means of legitimate, that is allegedly legitimate, violence." [52] The consensus is very

[51] Georges Sorel, *Reflections on Violence*, "Introduction to the First Publication" (1906), New York, 1961, p. 60.

[52] *The Power Elite*, New York, 1956, p. 171; Max Weber in the first paragraphs of *Politics as a Vocation* (1921). Weber seems to have been aware of his agreement with the Left. He quotes in the context Trotsky's remark in Brest-Litovsk, "Every state is based on violence," and adds, "This is indeed true."

strange; for to equate political power with "the organization of violence" makes sense only if one follows Marx's
estimate of the state as an instrument of oppression in the
hands of the ruling class. Let us therefore turn to authors
who do not believe that the body politic and its laws and
institutions are merely coercive superstructures, secondary
manifestations of some underlying forces. Let us turn, for
instance, to Bertrand de Jouvenel, whose book *Power* is
perhaps the most prestigious and, anyway, the most interesting recent treatise on the subject. "To him," he writes,
"who contemplates the unfolding of the ages war presents
itself as an activity of States *which pertains to their essence*." [53] This may prompt us to ask whether the end of
warfare, then, would mean the end of states. Would the
disappearance of violence in relationships between states
spell the end of power?

The answer, it seems, will depend on what we understand by power. And power, it turns out, is an instrument
of rule, while rule, we are told, owes its existence to "the
instinct of domination." [54] We are immediately reminded
of what Sartre said about violence when we read in
Jouvenel that "a man feels himself more of a man when he
is imposing himself and making others the instruments of
his will," which gives him "incomparable pleasure." [55]
"Power," said Voltaire, "consists in making others act as I
choose"; it is present wherever I have the chance "to as-
sert my own will against the resistance" of others, said Max
Weber, reminding us of Clausewitz's definition of war as
"an act of violence to compel the opponent to do as we
wish." The word, we are told by Strausz-Hupé, signifies

[53] *Power: The Natural History of Its Growth* (1945), London, 1952,
p. 122.

[54] *Ibidem*, p. 93.

[55] *Ibidem*, p. 110.

36

"the power of man over man." [56] To go back to Jouvenel: "To command and to be obeyed: without that, there is no Power—with it no other attribute is needed for it to be. . . . The thing without which it cannot be: that essence is command." [57] If the essence of power is the effectiveness of command, then there is no greater power than that which grows out of the barrel of a gun, and it would be difficult to say in "which way the order given by a policeman is different from that given by a gunman." (I am quoting from the important book *The Notion of the State,* by Alexander Passerin d'Entrèves, the only author I know who is aware of the importance of distinguishing between violence and power. "We have to decide whether and in what sense 'power' can be distinguished from 'force', to ascertain how the fact of using force according to law changes the quality of force itself and presents us with an entirely different picture of human relations," since "force, by the very fact of being qualified, ceases to be force." But even this distinction, by far the most sophisticated and thoughtful one in the literature, does not go

[56] See Karl von Clausewitz, *On War* (1832), New York, 1943, ch. 1; Robert Strausz-Hupé, *Power and Community,* New York, 1956, p. 4; the quotation from Max Weber: *"Macht bedeutet jede Chance, innerhalb einer sozialen Beziehung den eigenen Willen auch gegen Widerstand durchzusetzen,"* is drawn from Strausz-Hupé.

[57] I chose my examples at random, since it hardly matters to which author one turns. It is only occasionally that one hears a dissenting voice. Thus R. M. McIver states, "Coercive power is a criterion of the state, but not its essence. . . . It is true that there is no state, where there is no overwhelming force. . . . But the exercise of force does not make a state." (In *The Modern State,* London, 1926, pp. 222-225.) How strong the force of this tradition is can be seen in Rousseau's attempt to escape it. Looking for a government of no-rule, he finds nothing better than *"une forme d'association . . . par laquelle chacun s'unissant à tous n'obéisse pourtant qu'à lui-même."* The emphasis on obedience, and hence on command, is unchanged.

to the root of the matter. Power in Passerin d'Entrèves's understanding is "qualified" or "institutionalized force." In other words, while the authors quoted above define violence as the most flagrant manifestation of power, Passerin d'Entrèves defines power as a kind of mitigated violence. In the final analysis, it comes to the same.) [58] Should everybody from Right to Left, from Bertrand de Jouvenel to Mao Tse-tung agree on so basic a point in political philosophy as the nature of power?

In terms of our traditions of political thought, these definitions have much to recommend them. Not only do they derive from the old notion of absolute power that accompanied the rise of the sovereign European nation-state, whose earliest and still greatest spokesmen were Jean Bodin, in sixteenth-century France, and Thomas Hobbes, in seventeenth-century England; they also coincide with the terms used since Greek antiquity to define the forms of government as the rule of man over man—of one or the few in monarchy and oligarchy, of the best or the many in aristocracy and democracy. Today we ought to add the latest and perhaps most formidable form of such dominion: bureaucracy or the rule of an intricate system of bureaus in which no men, neither one nor the best, neither the few nor the many, can be held responsible, and which could be properly called rule by Nobody. (If, in accord with traditional political thought, we identify tyranny as government that is not held to give account of itself, rule by Nobody is clearly the most tyrannical of all, since there is no one left who could even be asked to answer for what

[58] *The Notion of the State, An Introduction to Political Theory* was first published in Italian in 1962. The English version is no mere translation; written by the author himself, it is the definitive edition and appeared in Oxford in 1967. For the quotations, see pp. 64, 70, and 105.

is being done. It is this state of affairs, making it impossible to localize responsibility and to identify the enemy, that is among the most potent causes of the current worldwide rebellious unrest, its chaotic nature, and its dangerous tendency to get out of control and to run amuck.)

Moreover, this ancient vocabulary was strangely confirmed and fortified by the addition of the Hebrew-Christian tradition and its "imperative conception of law." This concept was not invented by the "political realists" but was, rather, the result of a much earlier, almost automatic generalization of God's "Commandments," according to which "the simple relation of command and obedience" indeed sufficed to identify the essence of law.[59] Finally, more modern scientific and philosophical convictions concerning man's nature have further strengthened these legal and political traditions. The many recent discoveries of an inborn instinct of domination and an innate aggressiveness in the human animal were preceded by very similar philosophic statements. According to John Stuart Mill, "the first lesson of civilization [is] that of obedience," and he speaks of "the two states of the inclinations . . . one the desire to exercise power over others; the other . . . disinclination to have power exercised over themselves." [60] If we would trust our own experiences in these matters, we should know that the instinct of submission, an ardent desire to obey and be ruled by some strong man, is at least as prominent in human psychology as the will to power, and, politically, perhaps more relevant. The old adage "How fit he is to sway / That can so well obey," some version of which seems to have been

[59] *Ibidem,* p. 129.

[60] *Considerations on Representative Government* (1861), Liberal Arts Library, pp. 59 and 65.

known to all centuries and all nations,[61] may point to a psychological truth: namely, that the will to power and the will to submission are interconnected. "Ready submission to tyranny," to use Mill once more, is by no means always caused by "extreme passiveness." Conversely, a strong disinclination to obey is often accompanied by an equally strong disinclination to dominate and command. Historically speaking, the ancient institution of slave economy would be inexplicable on the grounds of Mill's psychology. Its express purpose was to liberate citizens from the burden of household affairs and to permit them to enter the public life of the community, where all were equals; if it were true that nothing is sweeter than to give commands and to rule others, the master would never have left his household.

However, there exists another tradition and another vocabulary no less old and time-honored. When the Athenian city-state called its constitution an isonomy, or the Romans spoke of the *civitas* as their form of government, they had in mind a concept of power and law whose essence did not rely on the command-obedience relationship and which did not identify power and rule or law and command. It was to these examples that the men of the eighteenth-century revolutions turned when they ransacked the archives of antiquity and constituted a form of government, a republic, where the rule of law, resting on the power of the people, would put an end to the rule of man over man, which they thought was a "government fit for slaves." They too, unhappily, still talked about obedience—obedience to laws instead of men; but what they actually meant was support of the laws to which the

[61] John M. Wallace, *Destiny His Choice: The Loyalism of Andrew Marvell*, Cambridge, 1968, pp. 88-89. I owe this reference to the kind attention of Gregory DesJardins.

citizenry had given its consent.[62] Such support is never unquestioning, and as far as reliability is concerned it cannot match the indeed "unquestioning obedience" that an act of violence can exact—the obedience every criminal can count on when he snatches my pocketbook with the help of a knife or robs a bank with the help of a gun. It is the people's support that lends power to the institutions of a country, and this support is but the continuation of the consent that brought the laws into existence to begin with. Under conditions of representative government the people are supposed to rule those who govern them. All political institutions are manifestations and materializations of power; they petrify and decay as soon as the living power of the people ceases to uphold them. This is what Madison meant when he said "all governments rest on opinion," a word no less true for the various forms of monarchy than for democracies. ("To suppose that majority rule functions only in democracy is a fantastic illusion," as Jouvenel points out: "The king, who is but one solitary individual, stands far more in need of the general support of Society than any other form of government." [63] Even the tyrant, the One who rules against all, needs helpers in the business of violence, though their number may be rather restricted.) However, the strength of opinion, that is, the power of the government, depends on numbers; it is "in proportion to the number with which it is associated," [64] and tyranny, as Montesquieu discovered, is therefore the most violent and least powerful of forms of government. Indeed one of the most obvious distinctions between power and violence is that

[62] See appendix XI, p. 97.

[63] *Op. cit.*, p. 98.

[64] *The Federalist*. No. 49.

power always stands in need of numbers, whereas violence up to a point can manage without them because it relies on implements. A legally unrestricted majority rule, that is, a democracy without a constitution, can be very formidable in the suppression of the rights of minorities and very effective in the suffocation of dissent without any use of violence. But that does not mean that violence and power are the same.

The extreme form of power is All against One, the extreme form of violence is One against All. And this latter is never possible without instruments. To claim, as is often done, that a tiny unarmed minority has successfully, by means of violence—shouting, kicking up a row, et cetera—disrupted large lecture classes whose overwhelming majority had voted for normal instruction procedures is therefore very misleading. (In a recent case at some German university there was even one lonely "dissenter" among several hundred students who could claim such a strange victory.) What actually happens in such cases is something much more serious: the majority clearly refuses to use its power and overpower the disrupters; the academic processes break down because no one is willing to raise more than a voting finger for the *status quo*. What the universities are up against is the "immense negative unity" of which Stephen Spender speaks in another context. All of which proves only that a minority can have a much greater potential power than one would expect by counting noses in public-opinion polls. The merely onlooking majority, amused by the spectacle of a shouting match between student and professor, is in fact already the latent ally of the minority. (One need only imagine what would have happened had one or a few unarmed Jews in pre-Hitler Germany tried to disrupt the lecture of an anti-Semitic professor in order to understand the absurdity of the talk about the small "minorities of militants.")

It is, I think, a rather sad reflection on the present state of political science that our terminology does not distinguish among such key words as "power," "strength," "force," "authority," and, finally, "violence"—all of which refer to distinct, different phenomena and would hardly exist unless they did. (In the words of d'Entrèves, "might, power, authority: these are all words to whose exact implications no great weight is attached in current speech; even the greatest thinkers sometimes use them at random. Yet it is fair to presume that they refer to different properties, and their meaning should therefore be carefully assessed and examined. . . . The correct use of these words is a question not only of logical grammar, but of historical perspective.") [65] To use them as synonyms not only indicates a certain deafness to linguistic meanings, which would be serious enough, but it has also resulted in a kind of blindness to the realities they correspond to. In such a situation it is always tempting to introduce new definitions, but—though I shall briefly yield to temptation—what is involved is not simply a matter of careless speech. Behind the apparent confusion is a firm conviction in whose light all distinctions would be, at best, of minor importance: the conviction that the most crucial political issue is, and always has been, the question of Who rules Whom? Power, strength, force, authority, violence—these are but words to indicate the means by which man rules over man; they are held to be synonyms because they have the same function. It is only after one

[65] *Op. cit.*, p. 7. Cf. also p. 171, where, discussing the exact meaning of the words "nation" and "nationality," he rightly insists that "the only competent guides in the jungle of so many different meanings are the linguists and the historians. It is to them that we must turn for help." And in distinguishing authority and power, he turns to Cicero's *potestas in populo, auctoritas in senatu*.

ceases to reduce public affairs to the business of dominion that the original data in the realm of human affairs will appear, or, rather, reappear, in their authentic diversity.

These data, in our context, may be enumerated as follows:

Power corresponds to the human ability not just to act but to act in concert. Power is never the property of an individual; it belongs to a group and remains in existence only so long as the group keeps together. When we say of somebody that he is "in power" we actually refer to his being empowered by a certain number of people to act in their name. The moment the group, from which the power originated to begin with (*potestas in populo,* without a people or group there is no power), disappears, "his power" also vanishes. In current usage, when we speak of a "powerful man" or a "powerful personality," we already use the word "power" metaphorically; what we refer to without metaphor is "strength."

Strength unequivocally designates something in the singular, an individual entity; it is the property inherent in an object or person and belongs to its character, which may prove itself in relation to other things or persons, but is essentially independent of them. The strength of even the strongest individual can always be overpowered by the many, who often will combine for no other purpose than to ruin strength precisely because of its peculiar independence. The almost instinctive hostility of the many toward the one has always, from Plato to Nietzsche, been ascribed to resentment, to the envy of the weak for the strong, but this psychological interpretation misses the point. It is in the nature of a group and its power to turn against independence, the property of individual strength.

Force, which we often use in daily speech as a synonym for violence, especially if violence serves as a means of coercion, should be reserved, in terminological language, for the "forces of nature" or the "force of circumstances" (*la force des choses*), that is, to indicate the energy released by physical or social movements.

Authority, relating to the most elusive of these phenomena and therefore, as a term, most frequently abused,[66] can be vested in persons—there is such a thing as personal authority, as, for instance, in the relation between parent and child, between teacher and pupil—or it can be vested in offices, as, for instance, in the Roman senate (*auctoritas in senatu*) or in the hierarchical offices of the Church (a priest can grant valid absolution even though he is drunk). Its hallmark is unquestioning recognition by those who are asked to obey; neither coercion nor persuasion is needed. (A father can lose his authority either by beating his child or by starting to argue with him, that is, either by behaving to him like a tyrant or by treating him as an equal.) To remain in authority requires respect for the person or the office. The greatest enemy of authority, therefore, is contempt, and the surest way to undermine it is laughter.[67]

[66] There is such a thing as authoritarian government, but it certainly has nothing in common with tyranny, dictatorship, or totalitarian rule. For a discussion of the historical background and political significance of the term, see my "What is Authority?" in *Between Past and Future: Exercises in Political Thought,* New York, 1968, and Part I of Karl-Heinz Lübke's valuable study, *Auctoritas bei Augustin,* Stuttgart, 1968, with extensive bibliography.

[67] Wolin and Schaar, in *op. cit.,* are entirely right: "The rules are being broken because University authorities, administrators and faculty alike, have lost the respect of many of the students." They then conclude, "When authority leaves, power enters." This too is true, but, I am afraid, not quite in the sense they meant it. What entered first at Berkeley was student power, obviously the strongest power on every campus simply because of the students' superior numbers. It was in order to break this power that authorities resorted to violence, and it is precisely because the university is essentially an institution based on authority, and therefore in need

45

Violence, finally, as I have said, is distinguished by its instrumental character. Phenomenologically, it is close to strength, since the implements of violence, like all other tools, are designed and used for the purpose of multiplying natural strength until, in the last stage of their development, they can substitute for it.

It is perhaps not superfluous to add that these distinctions, though by no means arbitrary, hardly ever correspond to watertight compartments in the real world, from which nevertheless they are drawn. Thus institutionalized power in organized communities often appears in the guise of authority, demanding instant, unquestioning recognition; no society could function without it. (A small, and still isolated, incident in New York shows what can happen if authentic authority in social relations has broken down to the point where it cannot work any longer even in its derivative, purely functional form. A minor mishap in the subway system—the doors on a train failed to operate—turned into a serious shutdown on the line lasting four hours and involving more than fifty thousand passengers, because when the transit authorities asked the passengers to leave the defective train, they simply refused.) [68] Moreover, nothing, as we shall see, is

---

of respect, that it finds it so difficult to deal with power in nonviolent terms. The university today calls upon the police for protection exactly as the Catholic church used to do before the separation of state and church forced it to rely on authority alone. It is perhaps more than an oddity that the severest crisis of the church as an institution should coincide with the severest crisis in the history of the university, the only secular institution still based on authority. Both may indeed be ascribed to "the progressing explosion of the atom 'obedience' whose stability was allegedly eternal," as Heinrich Böll remarked of the crisis in the churches. See "Es wird immer später," in *Antwort an Sacharow,* Zürich, 1969.

[68] See the New York *Times,* January 4, 1969, pp. 1 and 29.

more common than the combination of violence and power, nothing less frequent than to find them in their pure and therefore extreme form. From this, it does not follow that authority, power, and violence are all the same.

Still it must be admitted that it is particularly tempting to think of power in terms of command and obedience, and hence to equate power with violence, in a discussion of what actually is only one of power's special cases—namely, the power of government. Since in foreign relations as well as domestic affairs violence appears as a last resort to keep the power structure intact against individual challengers—the foreign enemy, the native criminal—it looks indeed as though violence were the prerequisite of power and power nothing but a façade, the velvet glove which either conceals the iron hand or will turn out to belong to a paper tiger. On closer inspection, though, this notion loses much of its plausibility. For our purpose, the gap between theory and reality is perhaps best illustrated by the phenomenon of revolution.

Since the beginning of the century theoreticians of revolution have told us that the chances of revolution have significantly decreased in proportion to the increased destructive capacities of weapons at the unique disposition of governments.[69] The history of the last seventy years,

[69] Thus Franz Borkenau, reflecting on the defeat of the Spanish revolution, states: "In this tremendous contrast with previous revolutions one fact is reflected. Before these latter years, counter-revolution usually depended upon the support of reactionary powers, which were technically and intellectually inferior to the forces of revolution. This has changed with the advent of fascism. Now, every revolution is likely to meet the attack of the most modern, most efficient, most ruthless machinery yet in existence. It means that the age of revolutions free to evolve according to their own laws is over." This was written more than thirty years ago (*The Spanish*

47

with its extraordinary record of successful and unsuccessful revolutions, tells a different story. Were people mad who even tried against such overwhelming odds? And, leaving out instances of full success, how can even a temporary success be explained? The fact is that the gap between state-owned means of violence and what people can muster by themselves—from beer bottles to Molotov cocktails and guns—has always been so enormous that technical improvements make hardly any difference. Textbook instructions on "how to make a revolution" in a step-by-step progression from dissent to conspiracy, from resistance to armed uprising, are all based on the mistaken notion that revolutions are "made." In a contest of violence against violence the superiority of the government has always been absolute; but this superiority lasts only as long as the power structure of the government is intact—that is, as long as commands are obeyed and the army or police forces are prepared to use their weapons. When this is no longer the case, the situation changes abruptly. Not only is the rebellion not put down, but the arms themselves change hands—sometimes, as in the Hungarian revolution, within a few hours. (We should know about such things after all these years of futile fighting in Vietnam, where for a long time, before getting massive Russian aid, the National Liberation Front fought us with weapons that were made in the United States.) Only after this has happened, when the disintegration of the government in power has permitted the rebels to arm themselves, can one speak of an "armed uprising," which often does not

---

Cockpit, London, 1937; Ann Arbor, 1963, pp. 288-289) and is now quoted with approval by Chomsky (*op. cit.*, p. 310). He believes that American and French intervention in the civil war in Vietnam proves Borkenau's prediction accurate, "with substitution of 'liberal imperialism' for 'fascism.'" I think that this example is rather apt to prove the opposite.

take place at all or occurs when it is no longer necessary. Where commands are no longer obeyed, the means of violence are of no use; and the question of this obedience is not decided by the command-obedience relation but by opinion, and, of course, by the number of those who share it. Everything depends on the power behind the violence. The sudden dramatic breakdown of power that ushers in revolutions reveals in a flash how civil obedience —to laws, to rulers, to institutions—is but the outward manifestation of support and consent.

Where power has disintegrated, revolutions are possible but not necessary. We know of many instances when utterly impotent regimes were permitted to continue in existence for long periods of time—either because there was no one to test their strength and reveal their weakness or because they were lucky enough not to be engaged in war and suffer defeat. Disintegration often becomes manifest only in direct confrontation; and even then, when power is already in the street, some group of men prepared for such an eventuality is needed to pick it up and assume responsibility. We have recently witnessed how it did not take more than the relatively harmless, essentially nonviolent French students' rebellion to reveal the vulnerability of the whole political system, which rapidly disintegrated before the astonished eyes of the young rebels. Unknowingly they had tested it; they intended only to challenge the ossified university system, and down came the system of governmental power, together with that of the huge party bureaucracies—"*une sorte de désintégration de toutes les hiérarchies.*" [70] It was a textbook case of a revolutionary situation [71] that did not develop into a revo-

[70] Raymond Aron, *La Révolution Introuvable,* 1968, p. 41.

[71] Stephen Spender, *op. cit.,* p. 56, disagrees: "What was so much more apparent than the revolutionary situation [was] the nonrevolutionary one." It may be "difficult to think of a revolution

lution because there was nobody, least of all the students, prepared to seize power and the responsibility that goes with it. Nobody except, of course, de Gaulle. Nothing was more characteristic of the seriousness of the situation than his appeal to the army, his journey to see Massu and the generals in Germany, a walk to Canossa, if there ever was one, in view of what had happened only a few years before. But what he sought and received was support, not obedience, and the means were not commands but concessions.[72] If commands had been enough, he would never have had to leave Paris.

No government exclusively based on the means of violence has ever existed. Even the totalitarian ruler, whose chief instrument of rule is torture, needs a power basis—the secret police and its net of informers. Only the development of robot soldiers, which, as previously mentioned, would eliminate the human factor completely and, conceivably, permit one man with a push button to destroy whomever he pleased, could change this fundamental ascendancy of power over violence. Even the most despotic domination we know of, the rule of master over slaves, who always outnumbered him, did not rest on superior means of coercion as such, but on a superior organization of power—that is, on the organized solidarity of the masters.[73] Single men without others to support them never

---

taking place when . . . everyone looks particularly good humoured," but this is what usually happens in the beginning of revolutions—during the early great ecstasy of fraternity.

[72] See appendix XII, p. 98.

[73] In ancient Greece, such an organization of power was the polis, whose chief merit, according to Xenophon, was that it permitted the "citizens to act as bodyguards to one another against slaves and criminals so that none of the citizens may die a violent death." (*Hiero*, IV, 3.)

have enough power to use violence successfully. Hence, in domestic affairs, violence functions as the last resort of power against criminals or rebels—that is, against single individuals who, as it were, refuse to be overpowered by the consensus of the majority. And as for actual warfare, we have seen in Vietnam how an enormous superiority in the means of violence can become helpless if confronted with an ill-equipped but well-organized opponent who is much more powerful. This lesson, to be sure, was there to be learned from the history of guerrilla warfare, which is at least as old as the defeat in Spain of Napoleon's still-unvanquished army.

To switch for a moment to conceptual language: Power is indeed of the essence of all government, but violence is not. Violence is by nature instrumental; like all means, it always stands in need of guidance and justification through the end it pursues. And what needs justification by something else cannot be the essence of anything. The end of war—end taken in its twofold meaning—is peace or victory; but to the question And what is the end of peace? there is no answer. Peace is an absolute, even though in recorded history periods of warfare have nearly always outlasted periods of peace. Power is in the same category; it is, as they say, "an end in itself." (This, of course, is not to deny that governments pursue policies and employ their power to achieve prescribed goals. But the power structure itself precedes and outlasts all aims, so that power, far from being the means to an end, is actually the very condition enabling a group of people to think and act in terms of the means-end category.) And since government is essentially organized and institutionalized power, the current question What is the end of government? does not make much sense either. The answer will be either question-begging—to enable men to live together—or danger-ously utopian—to promote happiness or to realize a classless society or some other nonpolitical ideal, which if tried out in earnest cannot but end in some kind of tyranny.

Power needs no justification, being inherent in the very existence of political communities; what it does need is legitimacy. The common treatment of these two words as synonyms is no less misleading and confusing than the current equation of obedience and support. Power springs up whenever people get together and act in concert, but it derives its legitimacy from the initial getting together rather than from any action that then may follow. Legitimacy, when challenged, bases itself on an appeal to the past, while justification relates to an end that lies in the future. Violence can be justifiable, but it never will be legitimate. Its justification loses in plausibility the farther its intended end recedes into the future. No one questions the use of violence in self-defense, because the danger is not only clear but also present, and the end justifying the means is immediate.

Power and violence, though they are distinct phenomena, usually appear together. Wherever they are combined, power, we have found, is the primary and predominant factor. The situation, however, is entirely different when we deal with them in their pure states—as, for instance, with foreign invasion and occupation. We saw that the current equation of violence with power rests on government's being understood as domination of man over man by means of violence. If a foreign conqueror is confronted by an impotent government and by a nation unused to the exercise of political power, it is easy for him to achieve such domination. In all other cases the difficulties are great indeed, and the occupying invader will try immediately to establish Quisling governments, that is, to find a native power base to support his dominion. The head-on clash between Russian tanks and the entirely

nonviolent resistance of the Czechoslovak people is a textbook case of a confrontation between violence and power
in their pure states. But while domination in such an
instance is difficult to achieve, it is not impossible. Violence, we must remember, does not depend on numbers
or opinions, but on implements, and the implements of
violence, as I mentioned before, like all other tools, increase and multiply human strength. Those who oppose
violence with mere power will soon find that they are confronted not by men but by men's artifacts, whose inhumanity and destructive effectiveness increase in proportion to the distance separating the opponents. Violence
can always destroy power; out of the barrel of a gun grows
the most effective command, resulting in the most instant
and perfect obedience. What never can grow out of it is
power.

In a head-on clash between violence and power, the
outcome is hardly in doubt. If Gandhi's enormously
powerful and successful strategy of nonviolent resistance
had met with a different enemy—Stalin's Russia, Hitler's
Germany, even prewar Japan, instead of England—the
outcome would not have been decolonization, but
massacre and submission. However, England in India and
France in Algeria had good reasons for their restraint.
Rule by sheer violence comes into play where power is
being lost; it is precisely the shrinking power of the Russian government, internally and externally, that became
manifest in its "solution" of the Czechoslovak problem—
just as it was the shrinking power of European imperialism that became manifest in the alternative between decolonization and massacre. To substitute violence for
power can bring victory, but the price is very high; for it
is not only paid by the vanquished, it is also paid by the
victor in terms of his own power. This is especially true
when the victor happens to enjoy domestically the bless-

ings of constitutional government. Henry Steele Commager is entirely right: "If we subvert world order and destroy world peace we must inevitably subvert and destroy our own political institutions first." [74] The much-feared boomerang effect of the "government of subject races" (Lord Cromer) on the home government during the imperialist era meant that rule by violence in faraway lands would end by affecting the government of England, that the last "subject race" would be the English themselves. The recent gas attack on the campus at Berkeley, where not just tear gas but also another gas, "outlawed by the Geneva Convention and used by the Army to flush out guerrillas in Vietnam," was laid down while gas-masked Guardsmen stopped anybody and everybody "from fleeing the gassed area," is an excellent example of this "backlash" phenomenon. It has often been said that impotence breeds violence, and psychologically this is quite true, at least of persons possessing natural strength, moral or physical. Politically speaking, the point is that loss of power becomes a temptation to substitute violence for power—in 1968 during the Democratic convention in Chicago we could watch this process on television [75]—and that violence itself results in impotence. Where violence is no longer backed and restrained by power, the well-known reversal in reckoning with means and ends has taken place. The means, the means of destruction, now determine the end—with the consequence that the end will be the destruction of all power.

Nowhere is the self-defeating factor in the victory of violence over power more evident than in the use of terror to maintain domination, about whose weird suc-

[74] "Can We Limit Presidential Power?" in *The New Republic*, April 6, 1968.

[75] See appendix XIII, p. 98.

cesses and eventual failures we know perhaps more than any generation before us. Terror is not the same as violence; it is, rather, the form of government that comes into being when violence, having destroyed all power, does not abdicate but, on the contrary, remains in full control. It has often been noticed that the effectiveness of terror depends almost entirely on the degree of social atomization. Every kind of organized opposition must disappear before the full force of terror can be let loose. This atomization—an outrageously pale, academic word for the horror it implies—is maintained and intensified through the ubiquity of the informer, who can be literally omnipresent because he no longer is merely a professional agent in the pay of the police but potentially every person one comes into contact with. How such a fully developed police state is established and how it works—or, rather, how nothing works where it holds sway—can now be learned in Aleksandr I. Solzhenitsyn's *The First Circle*, which will probably remain one of the masterpieces of twentieth-century literature and certainly contains the best documentation on Stalin's regime in existence.[76] The decisive difference between totalitarian domination, based on terror, and tyrannies and dictatorships, established by violence, is that the former turns not only against its enemies but against its friends and supporters as well, being afraid of all power, even the power of its friends. The climax of terror is reached when the police state begins to devour its own children, when yesterday's executioner becomes today's victim. And this is also the moment when power disappears entirely. There exist now a great many plausible explanations for the de-Stalinization of Russia—none, I believe, so compelling as the realization by the Stalinist functionaries themselves that a continua-

[76] See appendix XIV, p. 99.

tion of the regime would lead, not to an insurrection, against which terror is indeed the best safeguard, but to paralysis of the whole country.

To sum up: politically speaking, it is insufficient to say that power and violence are not the same. Power and violence are opposites; where the one rules absolutely, the other is absent. Violence appears where power is in jeopardy, but left to its own course it ends in power's disappearance. This implies that it is not correct to think of the opposite of violence as nonviolence; to speak of nonviolent power is actually redundant. Violence can destroy power; it is utterly incapable of creating it Hegel's and Marx's great trust in the dialectial "power of negation," by virtue of which opposites do not destroy but smoothly develop into each other because contradictions promote and do not paralyze development, rests on a much older philosophical prejudice: that evil is no more than a privative *modus* of the good, that good can come out of evil; that, in short, evil is but a temporary manifestation of a still-hidden good. Such time-honored opinions have become dangerous. They are shared by many who have never heard of Hegel or Marx, for the simple reason that they inspire hope and dispel fear—a treacherous hope used to dispel legitimate fear. By this, I do not mean to equate violence with evil; I only want to stress that violence cannot be derived from its opposite, which is power, and that in order to understand it for what it is, we shall have to examine its roots and nature.

T O S P E A K about the nature and causes of violence in these terms must appear presumptuous at a moment when floods of foundation money are channeled into the various research projects of social scientists, when a deluge of books on the subject has already appeared, when eminent natural scientists—biologists, physiologists, ethologists, and zoologists—have joined in an all-out effort to solve the riddle of "aggressiveness" in human behavior, and even a brand-new science, called "polemology," has emerged. I have two excuses for trying nevertheless.

First, while I find much of the work of the zoologists fascinating, I fail to see how it can possibly apply to our problem. In order to know that people will fight for their homeland we hardly had to discover instincts of "group territorialism" in ants, fish, and apes; and in order to learn that overcrowding results in irritation and aggressiveness, we hardly needed to experiment with rats. One day spent in the slums of any big city should have sufficed. I am surprised and often delighted to see that some animals behave like men; I cannot see how this could either justify or condemn human behavior. I fail to understand why we are asked "to recognize that man behaves very much like a group territorial species," rather than the other way round—that certain animal species behave very much like men.[77] (Following Adolf Portmann, these new insights into animal behavior do not close the gap between man and animal; they only demonstrate that "much more of what we know of ourselves than we thought also occurs in animals.") [78] Why should we, after having "eliminated" all anthropomorphisms from animal psychology (whether we actually succeeded is another matter), now try to discover "how 'theriomorph' man is"? [79] Is it not obvious that anthropomorphism and theriomorphism in the behavioral sciences are but two sides of the same "error"? Moreover, if we define man as belonging to the animal kingdom, why should we ask him to take his standards of behavior from another animal species? The answer, I am afraid, is simple: It is easier to experiment with animals, and this not only for humanitarian reasons—that it is not nice to put us into cages; the trouble is men can cheat.

Second, the research results of both the social and the natural sciences tend to make violent behavior even more of a "natural" reaction than we would have been prepared to grant without them. Aggressiveness, defined as an instinctual drive, is said to play the same functional role

[77] Nikolas Tinbergen, "On War and Peace in Animals and Man," in *Science*, 160: 1411 (June 28, 1968).

[78] *Das Tier als soziales Wesen*, Zürich, 1953, pp. 237-238: *"Wer sich in die Tatsachen vertieft . . . der wird feststellen, dass die neuen Einblicke in die Differenziertheit tierischen Treibens uns zwingen, mit allzu einfachen Vorstellungen von höheren Tieren ganz entschieden aufzuräumen. Damit wird aber nicht etwa—wie zuweilen leichthin gefolgert wird—das Tierische dem Menschlichen immer mehr genähert. Es zeigt sich lediglich, dass viel mehr von dem, was wir von uns selbst kennen, auch beim Tier vorkommt."*

[79] See Erich von Holst, *Zur Verhaltensphysiologie bei Tieren und Menschen*, Gesammelte Abhandlungen, Vol. I, München, 1969, p. 239.

in the household of nature as the nutritive and sexual instincts in the life process of the individual and the species. But unlike these instincts, which are activated by compelling bodily needs on one side, by outside stimulants on the other, aggressive instincts in the animal kingdom seem to be independent of such provocation; on the contrary, lack of provocation apparently leads to instinct frustration, to "repressed" aggressiveness, which according to psychologists causes a damming up of "energy" whose eventual explosion will be all the more dangerous. (It is as though the *sensation* of hunger in man would increase with the decrease of hungry people.) [80] In this interpretation, violence without provocation is "natural"; if it has lost its *rationale,* basically its function in self-preservation, it becomes "irrational," and this is allegedly the reason why men can be more "beastly" than other animals. (In the literature we are constantly reminded of the generous behavior of wolves, who do not kill the defeated enemy.)

Quite apart from the misleading transposition of physical terms such as "energy" and "force" to biological and zoological data, where they do not make sense because they cannot be measured,[81] I fear there lurks behind these

[80] To counter the absurdity of this conclusion a distinction is made between endogenous, spontaneous instincts, for instance, aggression, and reactive drives such as hunger. But a distinction between spontaneity and reactivity makes no sense in a discussion of innate impulses. In the world of nature there is no spontaneity, properly speaking, and instincts or drives only manifest the highly complex way in which all living organisms, including man, are adapted to its processes.

[81] The hypothetical character of Konrad Lorenz's *On Aggression* (New York, 1966) is clarified in the interesting collection of essays on aggression and adaptation edited by Alexander Mitscherlich under the title *Bis hierher und nicht weiter. Ist die menschliche Aggression unbefriedbar?,* München, 1968.

newest "discoveries" the oldest definition of the nature of man—the definition of man as the *animal rationale,* according to which we are distinct from other animal species in nothing but the additional attribute of reason. Modern science, starting uncritically from this old assumption, has gone far in "proving" that men share all other properties with some species of the animal kingdom—except that the additional gift of "reason" makes man a more dangerous beast. It is the use of reason that makes us dangerously "irrational," because this reason is the property of an "aboriginally instinctual being." [82] The scientists know, of course, that it is man the toolmaker who has invented those long-range weapons that free him from the "natural" restraints we find in the animal kingdom, and that tool-making is a highly complex *mental* activity.[83] Hence science is called upon to cure us of the side effects of reason by manipulating and controlling our instincts, usually by finding harmless outlets for them after their "life-promoting function" has disappeared. The standard of behavior is again derived from other animal species, in which the function of the life instincts has not been destroyed through the intervention of human reason. And the specific distinction between man and beast is now, strictly speaking, no longer reason (the *lumen naturale* of the human animal) but science, the knowledge of these stand-

[82] von Holst, *op. cit.,* p. 283: *"Nicht, weil wir Verstandeswesen, sondern weil wir ausserdem ganz urtümliche Triebwesen sind, ist unser Dasein im Zeitalter der Technik gefährdet."*

[83] Long-range weapons, seen by the polemologists as having freed man's aggressive instincts to the point where the controls safeguarding the species do not work any longer (see Tinbergen, *op. cit.*), are taken by Otto Klineberg ("Fears of a Psychologist," in Calder, *op. cit.,* p. 208) rather as an indication "that personal aggressiveness played [no] important role as a motive for war." Soldiers, one would like to continue the argument, are not killers, and killers—those with "personal aggressiveness"—are probably not even good soldiers.

ards and the techniques applying them. According to this view, man acts irrationally and like a beast if he refuses to listen to the scientists or is ignorant of their latest findings. As against these theories and their implications, I shall argue in what follows that violence is neither beastly nor irrational—whether we understand these terms in the ordinary language of the humanists or in accordance with scientific theories.

That violence often springs from rage is a commonplace, and rage can indeed be irrational and pathological, but so can every other human affect. It is no doubt possible to create conditions under which men are dehumanized—such as concentration camps, torture, famine—but this does not mean that they become animal-like; and under such conditions, not rage and violence, but their conspicuous absence is the clearest sign of dehumanization. Rage is by no means an automatic reaction to misery and suffering as such; no one reacts with rage to an incurable disease or to an earthquake or, for that matter, to social conditions that seem to be unchangeable. Only where there is reason to suspect that conditions could be changed and are not does rage arise. Only when our sense of justice is offended do we react with rage, and this reaction by no means necessarily reflects personal injury, as is demonstrated by the whole history of revolution, where invariably members of the upper classes touched off and then led the rebellions of the oppressed and downtrodden. To resort to violence when confronted with outrageous events or conditions is enormously tempting because of its inherent immediacy and swiftness. To act with *deliberate* speed goes against the grain of rage and violence, but this does not make them irrational. On the contrary, in private as well as public life there are situations in which the very swiftness of a violent act may be the only appropriate remedy. The point is not that this permits us to let off steam—

which indeed can be equally well done by pounding the table or slamming the door. The point is that under certain circumstances violence—acting without argument or speech and without counting the consequences—is the only way to set the scales of justice right again. (Billy Budd, striking dead the man who bore false witness against him, is the classical example.) In this sense, rage and the violence that sometimes—not always—goes with it belong among the "natural" *human* emotions, and to cure man of them would mean nothing less than to dehumanize or emasculate him. That such acts, in which men take the law into their own hands for justice's sake, are in conflict with the constitutions of civilized communities is undeniable; but their antipolitical character, so manifest in Melville's great story, does not mean that they are inhuman or "merely" emotional.

Absence of emotions neither causes nor promotes rationality. "Detachment and equanimity" in view of "unbearable tragedy" can indeed be "terrifying," [84] namely, when they are not the result of control but an evident manifestation of incomprehension. In order to respond reasonably one must first of all be "moved," and the opposite of emotional is not "rational," whatever that may mean, but either the inability to be moved, usually a pathological phenomenon, or sentimentality, which is a perversion of feeling. Rage and violence turn irrational only when they are directed against substitutes, and this, I am afraid, is precisely what the psychiatrists and polemologists concerned with human aggressiveness recommend, and what corresponds, alas, to certain moods and unreflecting attitudes in society at large. We all know, for example, that

[84] I am paraphrasing a sentence of Noam Chomsky (*op. cit.*, p. 371), who is very good in exposing the "façade of toughmindedness and pseudoscience" and the intellectual "vacuity" behind it, especially in the debates about the war in Vietnam.

it has become rather fashionable among white liberals to react to Negro grievances with the cry, "We are all guilty," and Black Power has proved only too happy to take advantage of this "confession" to instigate an irrational "black rage." Where all are guilty, no one is; confessions of collective guilt are the best possible safeguard against the discovery of culprits, and the very magnitude of the crime the best excuse for doing nothing. In this particular instance, it is, in addition, a dangerous and obfuscating escalation of racism into some higher, less tangible regions. The real rift between black and white is not healed by being translated into an even less reconcilable conflict between collective innocence and collective guilt. "All white men are guilty" is not only dangerous nonsense but also racism in reverse, and it serves quite effectively to give the very real grievances and rational emotions of the Negro population an outlet into irrationality, an escape from reality.

Moreover, if we inquire historically into the causes likely to transform *engagés* into *enragés*, it is not injustice that ranks first, but hypocrisy. Its momentous role in the later stages of the French Revolution, when Robespierre's war on hypocrisy transformed the "despotism of liberty" into the Reign of Terror, is too well known to be discussed here; but it is important to remember that this war had been declared long before by the French moralists who saw in hypocrisy the vice of all vices and found it ruling supreme in "good society," which somewhat later was called bourgeois society. Not many authors of rank glorified violence for violence's sake; but these few—Sorel, Pareto, Fanon—were motivated by a much deeper hatred of bourgeois society and were led to a much more radical break with its moral standards than the conventional Left, which was chiefly inspired by compassion and a burning desire for justice. To tear the mask of hypocrisy from the face of the enemy, to unmask him and the devious machinations and manipulations that permit him to rule without using violent means, that is, to provoke action even at the risk of annihilation so that the truth may come out—these are still among the strongest motives in today's violence on the campuses and in the streets.[85] And this violence again is not irrational. Since men live in a world of appearances and, in their dealing with it, depend on manifestation, hypocrisy's conceits—as distinguished from expedient ruses, followed by disclosure in due time—cannot be met by so-called reasonable behavior. Words can be relied on only if one is sure that their function is to reveal and not to conceal. It is the semblance of rationality, much more than the interests behind it, that provokes rage. To use reason when reason is used as a trap is not "rational"; just as to use a gun in self-defense is not "irrational." This violent reaction against hypocrisy, however justifiable in its own terms, loses its *raison d'être* when it tries to develop a strategy of its own with specific goals; it becomes "irrational" the moment it is "rationalized," that is, the moment the re-action in the course of a contest turns into an action, and the hunt for suspects, accompanied by the psychological hunt for ulterior motives, begins.[86]

Although the effectiveness of violence, as I remarked before, does not depend on numbers—one machine gunner

---

[85] "If one reads the SDS publications one sees that they have frequently recommended provocations of the police as a strategy for 'unmasking' the violence of the authorities." Spender (*op. cit.*, p. 92) comments that this kind of violence "leads to doubletalk in which the provocateur is playing at one and the same time the role of assailant and victim." The war on hypocrisy harbors a number of great dangers, some of which I examined briefly in *On Revolution*, New York, 1963, pp. 91-101.

[86] See appendix XV, p. 99.

can hold hundreds of well-organized people at bay—nonetheless in collective violence its most dangerously attractive features come to the fore, and this by no means because there is safety in numbers. It is perfectly true that in military as well as revolutionary action "individualism is the first [value] to disappear"; [87] in its stead, we find a kind of group coherence which is more intensely felt and proves to be a much stronger, though less lasting, bond than all the varieties of friendship, civil or private.[88] To be sure, in all illegal enterprises, criminal or political, the group, for the sake of its own safety, will require "that each individual perform an irrevocable action" in order to burn his bridges to respectable society before he is admitted into the community of violence. But once a man is admitted, he will fall under the intoxicating spell of "the practice of violence [which] binds men together as a whole, since each individual forms a violent link in the great chain, a part of the great organism of violence which has surged upward." [89]

Fanon's words point to the well-known phenomenon of brotherhood on the battlefield, where the noblest, most selfless deeds are often daily occurrences. Of all equalizers, death seems to be the most potent, at least in the few extraordinary situations where it is permitted to play a political role. Death, whether faced in actual dying or in the inner awareness of one's own mortality, is perhaps the most antipolitical experience there is. It signifies that we shall disappear from the world of appearances and shall leave the company of our fellow-men, which are the condi-

[87] Fanon, *op. cit.* p. 47.

[88] J. Glenn Gray, *The Warriors* (New York, 1959; now available in paperback), is most perceptive and instructive on this point. It should be read by everyone interested in the practice of violence.

[89] Fanon, *op. cit.,* pp. 85 and 93, respectively.

tions of all politics. As far as human experience is concerned, death indicates an extreme of loneliness and impotence. But faced collectively and in action, death changes its countenance; now nothing seems more likely to intensify our vitality than its proximity. Something we are usually hardly aware of, namely, that our own death is accompanied by the potential immortality of the group we belong to and, in the final analysis, of the species, moves into the center of our experience. It is as though life itself, the immortal life of the species, nourished, as it were, by the sempiternal dying of its individual members, is "surging upward," is actualized in the practice of violence.

It would be wrong, I think, to speak here of mere sentiments. After all, one of the outstanding properties of the human condition is here finding an adequate experience. In our context, however, the point of the matter is that these experiences, whose elementary force is beyond doubt, have never found an institutional, political expression, and that death as an equalizer plays hardly any role in political philosophy, although human mortality—the fact that men are "mortals," as the Greeks used to say—was understood as the strongest motive for political action in prephilosophic political thought. It was the certainty of death that made men seek immortal fame in deed and word and that prompted them to establish a body politic which was potentially immortal. Hence, politics was precisely a means by which to escape from the equality before death into a distinction assuring some measure of deathlessness. (Hobbes is the only political philosopher in whose work death, in the form of fear of violent death, plays a crucial role. But it is not equality before death that is decisive for Hobbes; it is the equality of fear resulting from the equal ability to kill possessed by everyone that persuades men in the state of nature to bind themselves into a commonwealth.) At any event, no body politic I know of was ever founded on equality before death and its actualization in violence; the suicide squads in history, which were indeed organized on this principle and therefore often called themselves "brotherhoods," can hardly be counted among political organizations. But it is true that the strong fraternal sentiments collective violence engenders have misled many good people into the hope that a new community together with a "new man" will arise out of it. The hope is an illusion for the simple reason that no human relationship is more transitory than this kind of brotherhood, which can be actualized only under conditions of immediate danger to life and limb.

That, however, is but one side of the matter. Fanon concludes his praise of the practice of violence by remarking that in this kind of struggle the people realize "that life is an unending contest," that violence is an element of life. And does that not sound plausible? Have not men always equated death with "eternal rest," and does it not follow that where we have life we have struggle and unrest? Is not quiet a clear manifestation of lifelessness or decay? Is not violent action a prerogative of the young—those who presumably are fully alive? Therefore are not praise of life and praise of violence the same? Sorel, at any rate, thought along these lines sixty years ago. Before Spengler, he predicted the "Decline of the Occident," having observed clear signs of abatement in the European class struggle. The bourgeoisie, he argued, had lost the "energy" to play its role in the class struggle; only if the proletariat could be persuaded to use violence in order to reaffirm class distinctions and awaken the fighting spirit of the bourgeoisie could Europe be saved.[90]

Hence, long before Konrad Lorenz discovered the life-

[90] Sorel, *op. cit.*, chapter 2, "On Violence and the Decadence of the Middle Classes."

promoting function of aggression in the animal kingdom, violence was praised as a manifestation of the life force and specifically of its creativity. Sorel, inspired by Bergson's *élan vital*, aimed at a philosophy of creativity designed for "producers" and polemically directed against the consumer society and its intellectuals; both groups, he felt, were parasites. The image of the bourgeois—peaceful, complacent, hypocritical, bent on pleasure, without will to power, a late product of capitalism rather than its representative—and the image of the intellectual, whose theories are "constructions" instead of "expressions of the will," [91] are hopefully counterbalanced in his work by the image of the worker. Sorel sees the worker as the "producer," who will create the new "moral qualities, which are necessary to improve production," destroy "the Parliaments [which] are as packed as shareholders' meetings," [92] and oppose to "the image of Progress . . . the image of total catastrophe," when "a kind of irresistible wave will pass over the old civilization." [93] The new values turn out to be not very new. They are a sense of honor, desire for fame and glory, the spirit of fighting without hatred and "without the spirit of revenge," and indifference to material advantages. Still, they are indeed the very virtues that were conspicuously absent from bourgeois society.[94] "Social war, by making an appeal to the honor which develops so naturally in all organized armies, can eliminate those evil feelings against which morality would remain

[91] *Ibidem*, "Introduction, Letter to Daniel Halevy," iv.

[92] *Ibidem*, chapter 7, "The Ethics of the Producers," I.

[93] *Ibidem*, chapter 4, "The Proletarian Strike," II.

[94] *Ibidem*; see especially chapter 5, III, and chapter 3, "Prejudices against Violence," III.

powerless. If this were the only reason . . . this reason alone would, it seems to me, be decisive in favor of the apologists for violence." [95]

Much can be learned from Sorel about the motives that prompt men to glorify violence in the abstract, and even more from his more gifted Italian contemporary, also of French formation, Vilfredo Pareto. Fanon, who had an infinitely greater intimacy with the practice of violence than either, was greatly influenced by Sorel and used his categories even when his own experiences spoke clearly against them.[96] The decisive experience that persuaded Sorel as well as Pareto to stress the factor of violence in revolutions was the Dreyfus Affair in France, when, in the words of Pareto, they were "amazed to see [the Dreyfusards] employing against their opponents the same vil-

---

[95] *Ibidem,* Appendix 2, "Apology for Violence."

[96] This has recently been stressed by Barbara Deming in her plea for nonviolent action—"On Revolution and Equilibrium," in *Revolution: Violent and Nonviolent,* reprinted from *Liberation,* February, 1968. She says about Fanon, on p. 3: "It is my conviction that he can be quoted as well to plead for nonviolence. . . . Every time you find the word 'violence' in his pages, substitute for it the phrase 'radical and uncompromising action.' I contend that with the exception of a very few passages this substitution can be made, and that the action he calls for could just as well be nonviolent action." Even more important for my purposes: Miss Deming also tries to distinguish clearly between power and violence, and she recognizes that "nonviolent disruption" means "to exert force. . . . It resorts even to what can only be called physical force" (p. 6). However, she curiously underestimates the effect of this force of disruption, which stops short only of physical injury, when she says, "the human rights of the adversary are respected" (p. 7). Only the opponent's right to life, but none of the other human rights, is actually respected. The same is of course true for those who advocate "violence against things" as opposed to "violence against persons."

71

lainous methods that they had themselves denounced." [97]
At that juncture they discovered what we call today the
Establishment and what earlier was called the System, and
it was this discovery that made them turn to the praise of
violent action and made Pareto, for his part, despair of the
working class. (Pareto understood that the rapid integration of the workers into the social and political body of
the nation actually amounted to "an alliance of bourgeoisie
and working people," to the "embourgeoisement" of the
workers, which then, according to him, gave rise to a new
system, which he called "Pluto-democracy"—a mixed form
of government, plutocracy being the bourgeois regime and
democracy the regime of the workers.) The reason Sorel
held on to his Marxist faith in the working class was
that the workers were the "producers," the only creative
element in society, those who, according to Marx, were
bound to liberate the productive forces of mankind; the
trouble was only that as soon as the workers had reached
a satisfactory level of working and living conditions, they
stubbornly refused to remain proletarians and play their
revolutionary role.

Something else, however, which became fully manifest
only in the decades after Sorel's and Pareto's death, was
incomparably more disastrous to this view. The enormous
growth of productivity in the modern world was by no
means due to an increase in the workers' productivity, but
exclusively the development of technology, and this depended neither on the working class nor on the bourgeoisie, but on the scientists. The "intellectuals," much de-
spised by Sorel and Pareto, suddenly ceased to be a
marginal social group and emerged as a new elite, whose

[97] Quoted from S. E. Finer's instructive essay "Pareto and Pluto-
Democracy: The Retreat to Galapagos," in *The American Political
Science Review*, June, 1968.

work, having changed the conditions of human life almost beyond recognition in a few decades, has remained essential for the functioning of society. There are many reasons why this new group has not, or not yet, developed into a power elite, but there is indeed every reason to believe with Daniel Bell that "not only the best talents, but eventually the entire complex of social prestige and social status, will be rooted in the intellectual and scientific communities." [98] Its members are more dispersed and less bound by clear interests than groups in the old class system; hence, they have no drive to organize themselves and lack experience in all matters pertaining to power. Also, being much more closely bound to cultural traditions, of which the revolutionary tradition is one, they cling with greater tenacity to categories of the past that prevent them from understanding the present and their own role in it. It is often touching to watch with what nostalgic sentiments the most rebellious of our students expect the "true" revolutionary impetus to come from those groups in society that denounce them the more vehemently the more they have to lose by anything that could disturb the smooth functioning of the consumer society. For better or worse—and I think there is every reason to be fearful as well as hopeful—the really new and potentially revolutionary class in society will consist of intellectuals, and their potential power, as yet unrealized, is very great, perhaps too great for the good of mankind. [99] But these are speculations.

However that may be, in this context we are chiefly interested in the strange revival of the life philosophies of Bergson and Nietzsche in their Sorelian version. We all

[98] "Notes on the Post-Industrial Society," *The Public Interest,* No. 6, 1967.

[99] See appendix XVI, p. 100.

know to what extent this old combination of violence, life, and creativity figures in the rebellious state of mind of the present generation. No doubt the emphasis on the sheer factuality of living, and hence on love-making as life's most glorious manifestation, is a response to the real possibility of constructing a doomsday machine and destroying all life on earth. But the categories in which the new glorifiers of life understand themselves are not new. To see the productivity of society in the image of life's "creativity" is at least as old as Marx, to believe in violence as a life-promoting force is at least as old as Nietzsche, and to think of creativity as man's highest good is at least as old as Bergson.

And this seemingly so novel biological justification of violence is again closely connected with the most pernicious elements in our oldest traditions of political thought. According to the traditional concept of power, equated, as we saw, with violence, power is expansionist by nature. It "has an inner urge to grow," it is creative because "the instinct of growth is proper to it." [100] Just as in the realm of organic life everything either grows or declines and dies, so in the realm of human affairs power supposedly can sustain itself only through expansion; otherwise it shrinks and dies. "That which stops growing begins to rot," goes a Russian saying from the entourage of Catherine the Great. Kings, we are told, were killed "not because of their tyranny but because of their weakness. The people erect scaffolds, not as the moral punishment of despotism, but as the *biological* penalty for weakness" (my italics). Revolutions, therefore, were directed against the established powers "only to the outward view." Their true "effect was to give Power a new vigour and poise, and to pull down the obstacles which had long obstructed its de-

[100] Jouvenel, *op. cit.*, pp. 114 and 123, respectively.

velopment."[101] When Fanon speaks of the "creative madness" present in violent action, he is still thinking in this tradition.[102]

Nothing, in my opinion, could be theoretically more dangerous than the tradition of organic thought in political matters by which power and violence are interpreted in biological terms. As these terms are understood today, life and life's alleged creativity are their common denominator, so that violence is justified on the ground of creativity. The organic metaphors with which our entire present discussion of these matters, especially of the riots, is permeated —the notion of a "sick society," of which riots are symptoms, as fever is a symptom of disease—can only promote violence in the end. Thus the debate between those who propose violent means to restore "law and order" and those who propose nonviolent reforms begins to sound ominously like a discussion between two physicians who debate the relative advantages of surgical as opposed to medical treatment of their patient. The sicker the patient is supposed to be, the more likely that the surgeon will have the last word. Moreover, so long as we talk in nonpolitical, biological terms, the glorifiers of violence can appeal to the undeniable fact that in the household of nature destruction and creation are but two sides of the natural process, so that collective violent action, quite apart from its inherent attraction, may appear as natural a prerequisite for the collective life of mankind as the struggle for survival and violent death for continuing life in the animal kingdom.

The danger of being carried away by the deceptive plausibility of organic metaphors is particularly great where the racial issue is involved. Racism, white or black,

[101] *Ibidem,* pp. 187 and 188.

[102] Fanon, *op. cit.,* p. 95.

is fraught with violence by definition because it objects to natural organic facts—a white or black skin—which no persuasion or power could change; all one can do, when the chips are down, is to exterminate their bearers. Racism, as distinguished from race, is not a fact of life, but an ideology, and the deeds it leads to are not reflex actions, but deliberate acts based on pseudo-scientific theories. Violence in interracial struggle is always murderous, but it is not "irrational"; it is the logical and rational consequence of racism, by which I do not mean some rather vague prejudices on either side, but an explicit ideological system. Under the pressure of power, prejudices, as distinguished from both interests and ideologies, may yield—as we saw happen with the highly successful civil-rights movement, which was entirely nonviolent. ("By 1964 ... most Americans were convinced that subordination and, to a lesser degree, segregation were wrong.") [103] But while boycotts, sit-ins, and demonstrations were successful in eliminating discriminatory laws and ordinances in the South, they proved utter failures and became counterproductive when they encountered the social conditions in the large urban centers—the stark needs of the black ghettos on one side, the overriding interests of white lower-income groups in respect to housing and education on the other. All this mode of action could do, and indeed did, was to bring these conditions into the open, into the street, where the basic irreconcilability of interests was dangerously exposed.

But even today's violence, black riots, and the potential violence of the white backlash are not yet manifestations of racist ideologies and their murderous logic. (The riots,

[103] Robert M. Fogelson, "Violence as Protest," in *Urban Riots: Violence and Social Change,* Proceedings of the Academy of Political Science, Columbia University, 1968.

as has recently been stated, are "articulate protests against genuine grievances"; [104] indeed restraint and selectivity—or . . . rationality are certainly among [their] most crucial features." [105] And much the same is true for the backlash phenomena, which, contrary to all predictions, have not been characterized by violence up to now. It is the perfectly rational reaction of certain interest groups which furiously protest against being singled out to pay the full price for ill-designed integration policies whose consequences their authors can easily escape.) [106] The greatest danger comes from the other direction; since violence always needs justification, an escalation of the violence in the streets may bring about a truly racist ideology to justify it. Black racism, so blatantly evident in James Forman's "Manifesto" is probably more a reaction to the chaotic rioting of the last years than its cause. It could, of course, provoke a really violent white backlash, whose greatest danger would be the transformation of white prejudices into a full-fledged racist ideology for which "law and order" would indeed become a mere façade. In this still unlikely case, the climate of opinion in the country might deteriorate to the point where a majority of its citizens would be willing to pay the price of the invisible terror of a police state for law and order in the streets. What we have now, a kind of police backlash, quite brutal and highly visible, is nothing of the sort.

Behavior and arguments in interest conflicts are not notorious for their "rationality." Nothing, unfortunately, has so constantly been refuted by reality as the credo of "enlightened self-interest," in its literal version as well as

[104] *Ibidem.*

[105] *Ibidem.* See also the excellent article "Official Interpretation of Racial Riots" by Allan A. Silver in the same collection.

[106] See appendix XVII, p. 101.

in its more sophisticated Marxian variant. Some experience plus a little reflection teach, on the contrary, that it goes against the very nature of self-interest to be enlightened. To take as an example from everyday life the current interest conflict between tenant and landlord: enlightened interest would focus on a building fit for human habitation, but this interest is quite different from, and in most cases opposed to, the landlord's self-interest in high profit and the tenant's in low rent. The common answer of an arbiter, supposedly the spokesman of "enlightenment," namely, that *in the long run* the interest of the building is the *true* interest of both landlord and tenant, leaves out of account the time factor, which is of paramount importance for all concerned. Self-interest is interested in the self, and the self dies or moves out or sells the house; because of its changing condition, that is, ultimately because of the human condition of mortality, the self *qua* self cannot reckon in terms of long-range interest, i.e. the interest of a world that survives its inhabitants. Deterioration of the building is a matter of years; a rent increase or a temporarily lower profit rate are for today or for tomorrow. And something similar, *mutatis mutandis,* is of course true for labor-management conflicts and the like. Self-interest, when asked to yield to "true" interest—that is, the interest of the world as distinguished from that of the self—will always reply, Near is my shirt, but nearer is my skin. That may not be particularly reasonable, but it is quite realistic; it is the not very noble but adequate response to the time discrepancy between men's private lives and the altogether different life expectancy of the public world. To expect people, who have not the slightest notion of what the *res publica,* the public thing, is, to behave nonviolently and argue rationally in matters of interest is neither realistic nor reasonable.

Violence, being instrumental by nature, is rational to the extent that it is effective in reaching the end that must justify it. And since when we act we never know with any certainty the eventual consequences of what we are doing, violence can remain rational only if it pursues short-term goals. Violence does not promote causes, neither history nor revolution, neither progress nor reaction; but it can serve to dramatize grievances and bring them to public attention. As Conor Cruise O'Brien (in a debate on the legitimacy of violence in the Theatre of Ideas) once remarked, quoting William O'Brien, the nineteenth-century Irish agrarian and nationalist agitator: Sometimes "violence is the only way of ensuring a hearing for moderation." To ask the impossible in order to obtain the possible is not always counterproductive. And indeed, violence, contrary to what its prophets try to tell us, is more the weapon of reform than of revolution. France would not have received the most radical bill since Napoleon to change its antiquated education system if the French students had not rioted; if it had not been for the riots of the spring term, no one at Columbia University would have dreamed of accepting reforms; [107] and it is probably quite true that in West Germany the existence of "dissenting minorities is not even noticed unless they engage in provocation." [108] No doubt, "violence pays," but

[107] "At Columbia, before last year's uprising, for example, a report on student life and another on faculty housing had been gathering dust in the president's office," as Fred Hechinger reported in the New York *Times*, "The Week in Review" of May 4, 1969.

[108] Rudi Dutschke, as quoted in *Der Spiegel*, February 10, 1969, p. 27. Günter Grass, speaking in much the same vein after the attack on Dutschke in spring 1968, also stresses the relation between reforms and violence: "The youth protest movement has brought the fragility of our insufficiently established democracy into evidence.

the trouble is that it pays indiscriminately, for "soul courses" and instruction in Swahili as well as for real reforms. And since the tactics of violence and disruption make sense only for short-term goals, it is even more likely, as was recently the case in the United States, that the established power will yield to nonsensical and obviously damaging demands—such as admitting students without the necessary qualifications and instructing them in nonexistent subjects—if only such "reforms" can be made with comparative ease, than that violence will be effective with respect to the relatively long-term objective of structural change.[109] Moreover, the danger of violence, even if it moves consciously within a nonextremist framework of short-term goals, will always be that the means overwhelm the end. If goals are not achieved rapidly, the result will be not merely defeat but the introduction of the practice of violence into the whole body politic. Action is irreversible, and a return to the *status quo* in case of defeat is always unlikely. The practice of violence, like all action, changes the world, but the most probable change is to a more violent world.

In this it has been successful, but it is far from certain where this success will lead; either it will bring about long-overdue reforms . . . or . . . the uncertainty that has now been laid bare will provide false prophets with promising markets and free advertising." See "Violence Rehabilitated," in *Speak Out!*, New York, 1969.

[109] Another question, which we cannot discuss here, is to what an extent the whole university system is still capable of reforming itself. I think there is no general answer. Even though the student rebellion is a global phenomenon, the university systems themselves are by no means uniform and vary not only from country to country but often from institution to institution; all solutions of the problem must spring from, and correspond to, strictly local conditions. Thus, in some countries the university crisis may even broaden into a government crisis—as *Der Spiegel* (June 23, 1969) thought possible in discussing the German situation.

Finally—to come back to Sorel's and Pareto's earlier denunciation of the system as such—the greater the bureaucratization of public life, the greater will be the attraction of violence. In a fully developed bureaucracy there is nobody left with whom one can argue, to whom one can present grievances, on whom the pressures of power can be exerted. Bureaucracy is the form of government in which everybody is deprived of political freedom, of the power to act; for the rule by Nobody is not no-rule, and where all are equally powerless we have a tyranny without a tyrant. The crucial feature in the student rebellions around the world is that they are directed everywhere against the ruling bureaucracy. This explains what at first glance seems so disturbing—that the rebellions in the East demand precisely those freedoms of speech and thought that the young rebels in the West say they despise as irrelevant. On the level of ideologies, the whole thing is confusing; it is much less so if we start from the obvious fact that the huge party machines have succeeded everywhere in overruling the voice of the citizens, even in countries where freedom of speech and association is still intact. The dissenters and resisters in the East demand free speech and thought as the preliminary conditions for political action; the rebels in the West live under conditions where these preliminaries no longer open the channels for action, for the meaningful exercise of freedom. What matters to them is, indeed, *"Praxisentzug,"* the suspension of action, as Jens Litten, a German student, has aptly called it.[110] The transformation of government into administration, or of republics into bureaucracies, and the disastrous shrinkage of the public realm that went with it have a long and complicated history throughout the modern age; and this process has been considerably

[110] See appendix XVIII, p. 102.

accelerated during the last hundred years through the rise of party bureaucracies. (Seventy years ago Pareto recognized that "freedom . . . by which I mean the power to act shrinks every day, save for criminals, in the so-called free and democratic countries.) [111] What makes man a political being is his faculty of action; it enables him to get together with his peers, to act in concert, and to reach out for goals and enterprises that would never enter his mind, let alone the desires of his heart, had he not been given this gift—to embark on something new. Philosophically speaking, to act is the human answer to the condition of natality. Since we all come into the world by virtue of birth, as newcomers and beginnings, we are able to start something new; without the fact of birth we would not even know what novelty is, all "action" would be either mere behavior or preservation. No other faculty except language, neither reason nor consciousness, distinguishes us so radically from all animal species. To act and to begin are not the same, but they are closely interconnected.

None of the properties of creativity is adequately expressed in metaphors drawn from the life process. To beget and to give birth are no more creative than to die is annihilating; they are but different phases of the same, ever-recurring cycle in which all living things are held as though they were spellbound. Neither violence nor power is a natural phenomenon, that is, a manifestation of the life process; they belong to the political realm of human affairs whose essentially human quality is guaranteed by man's faculty of action, the ability to begin something new. And I think it can be shown that no other human ability has suffered to such an extent from the progress of the modern age, for progress, as we have come to understand it, means growth, the relentless process of more and

[111] Pareto, quoted from Finer, *op. cit.*

more, of bigger and bigger. The bigger a country becomes in terms of population, of objects, and of possessions, the greater will be the need for administration and with it the anonymous power of the administrators. Pavel Kohout, a Czech author, writing in the heyday of the Czechoslovakian experiment with freedom, defined a "free citizen" as a "Citizen-Co-ruler." He meant nothing more or less than the "participatory democracy" of which we have heard so much in recent years in the West. Kohout added that what the world today stands in greatest need of may well be "a new example" if "the next thousand years are not to become an era of supercivilized monkeys"—or, even worse, of "man turned into a chicken or a rat," ruled over by an "elite" that derives its power "from the wise counsels of . . . intellectual aides" who actually believe that men in think tanks are thinkers and that computers can think; "the counsels may turn out to be incredibly insidious and, instead of pursuing human objectives, may pursue completely abstract problems that had been transformed in an unforeseen manner in the artificial brain." [112]

This new example will hardly be set by the practice of violence, although I am inclined to think that much of the present glorification of violence is caused by severe frustration of the faculty of action in the modern world. It is simply true that riots in the ghettos and rebellions on the campuses make "people feel they are acting together in a way that they rarely can." [113] We do not know if these occurrences are the beginnings of something new—the "new example"—or the death pangs of a faculty that

[112] See Günter Grass and Pavel Kohout, *Briefe über die Grenze,* Hamburg, 1968, pp. 88 and 90, respectively; and Andrei D. Sakharov, *op. cit.*

[113] Herbert- J. Gans, "The Ghetto Rebellions and Urban Class Conflict," in *Urban Riots, op. cit.*

mankind is about to lose. As things stand today, when we see how the superpowers are bogged down under the monstrous weight of their own bigness, it looks as though the setting of a "new example" will have a chance, if at all, in a small country, or in small, well-defined sectors in the mass societies of the large powers.

The disintegration processes which have become so manifest in recent years—the decay of public services: schools, police, mail delivery, garbage collection, transportation, et cetera; the death rate on the highways and the traffic problems in the cities; the pollution of air and water—are the automatic results of the needs of mass societies that have become unmanageable. They are accompanied and often accelerated by the simultaneous decline of the various party systems, all of more or less recent origin and designed to serve the political needs of mass populations—in the West to make representative government possible when direct democracy would not do any longer because "the room will not hold all" (John Selden), and in the East to make absolute rule over vast territories more effective. Bigness is afflicted with vulnerability; cracks in the power structure of all but the small countries are opening and widening. And while no one can say with assurance where and when the breaking point has been reached, we can observe, almost measure, how strength and resiliency are insidiously destroyed, leaking, as it were, drop by drop from our institutions.

Moreover, there is the recent rise of a curious new brand of nationalism, usually understood as a swing to the Right, but more probably an indication of a growing, world-wide resentment against "bigness" as such. While national feelings formerly tended to unite various ethnic groups by focusing their political sentiments on the nation as a whole, we now watch how an ethnic "nationalism" begins to threaten with dissolution the oldest and best-established

84

nation-states. The Scots and the Welsh, the Bretons and the Provençals, ethnic groups whose successful assimilation had been the prerequisite for the rise of the nation-state and had seemed completely assured, are turning to separatism in rebellion against the centralized governments in London and Paris. And just when centralization, under the impact of bigness, turned out to be counterproductive in its own terms, this country, founded, according to the federal principle, on the division of powers and powerful so long as this division was respected, threw itself headlong, to the unanimous applause of all "progressive" forces, into the new, for America, experiment of centralized administration—the federal government overpowering state powers and executive power eroding congressional powers.[114] It is as though this most successful European colony wished to share the fate of the mother countries in their decline, repeating in great haste the very errors the framers of the Constitution had set out to correct and to eliminate.

Whatever the administrative advantages and disadvantages of centralization may be, its political result is always the same: monopolization of power causes the drying up or oozing away of all authentic power sources in the country. In the United States, based on a great plurality of powers and their mutual checks and balances, we are confronted not merely with the disintegration of power structures, but with power, seemingly still intact and free to manifest itself, losing its grip and becoming ineffective. To speak of the impotence of power is no longer a witty paradox. Senator Eugene McCarthy's crusade in 1968 "to test the system" brought popular resentment against imperialist adventures into the open, provided the link between the opposition in the Senate and

[114] See the important article of Henry Steele Commager, footnote 74.

that in the streets, enforced an at least temporary spectacular change in policy, and demonstrated how quickly the majority of the young rebels could become dealienated, jumping at this first opportunity not to abolish the system but to make it work again. And still, all this power could be crushed by the party bureaucracy, which, contrary to all traditions, preferred to lose the presidential election with an unpopular candidate who happened to be an *apparatchik*. (Something similar happened when Rockefeller lost the nomination to Nixon during the Republican convention.)

There are other examples to demonstrate the curious contradictions inherent in impotence of power. Because of the enormous effectiveness of teamwork in the sciences, which is perhaps the outstanding American contribution to modern science, we can control the most complicated processes with a precision that makes trips to the moon less dangerous than ordinary weekend excursions; but the allegedly "greatest power on earth" is helpless to end a war, clearly disastrous for all concerned, in one of the earth's smallest countries. It is as though we have fallen under a fairyland spell which permits us to do the "impossible" on the condition that we lose the capacity of doing the possible, to achieve fantastically extraordinary feats on the condition of no longer being able to attend properly to our everyday needs. If power has anything to do with the we-*will*-and-we-can, as distinguished from the mere we-can, then we have to admit that our power has become impotent. The progresses made by science have nothing to do with the I-will; they follow their own inexorable laws, compelling us to do whatever we can, regardless of consequences. Have the I-will and the I-can parted company? Was Valéry right when he said fifty years ago: *"On peut dire que tout ce que nous savons, c'est-à-dire tout ce que nous pouvons, a fini par s'opposer à ce que nous sommes*"? ("One can say that all we know, that is, all we have the power to do, has finally turned against what we are.")

Again, we do not know where these developments will lead us, but we know, or should know, that every decrease in power is an open invitation to violence—if only because those who hold power and feel it slipping from their hands, be they the government or be they the governed, have always found it difficult to resist the temptation to substitute violence for it.

Appendices

I, TO PAGE 13, NOTE 16

Professor B. C. Parekh, of Hull University, England, kindly drew my attention to the following passage in the section on Feuerbach from Marx's and Engels' *German Ideology* (1846), of which Engels later wrote: "The portion finished . . . only proves how incomplete at that time was our knowledge of economic history." "Both for the production on a mass scale of this communist consciousness, and for the success of the cause itself, the alteration of man [des Menschen] on a mass scale is necessary, an alteration which can only take place in a practical movement, a *revolution;* this revolution is necessary, therefore, not only because the ruling class cannot be overthrown in any other way, but also because the class *overthrowing* it can only in a revolution succeed in ridding itself of all the muck of ages and become fitted to found society anew." (Quoted from the edition by R. Pascal, New York, 1960, pp. xv and 69.) Even in these, as it were, pre-Marxist utterances, the distinction between Marx's and Sartre's positions is evident. Marx speaks of "the alteration of man on a mass scale," and of a "mass production of consciousness," not of the liberation of an individual through an isolated act of violence. (For the German text, see Marx/Engels *Gesamtausgabe,* 1932, I. Abteilung, vol. 5, pp. 59 f.)

II, TO PAGE 13, NOTE 17

The New Left's unconscious drifting away from Marxism has been duly noticed. See especially recent comments on the student movement by Leonard Schapiro in the *New York Review of Books* (December 5, 1968) and by Raymond Aron in *La Révolution Introuvable,* Paris, 1968. Both consider the new emphasis on violence to be a kind of backsliding either to pre-Marxian utopian socialism (Aron) or to the Russian anarchism of Nechaev and Bakunin (Schapiro), who "had much to say about the importance of violence as a factor of unity, as the binding force in a society or group, a

89

century before the same ideas emerged in the works of Jean-Paul Sartre and Frantz Fanon." Aron writes in the same vein: *"Les chantres de la révolution de mai croient dépasser le marxisme . . . ils oublient un siècle d'histoire"* (p. 14). To a non-Marxist such a reversion would of course hardly be an argument; but for Sartre, who, for instance, writes *"Un prétendu 'dépassement' du marxisme ne sera au pis qu'un retour au prémarxisme, au mieux que la redécouverte d'une pensée déjà contenue dans la philosophie qu'on a cru dépasser"* ("Question de Méthode" in *Critique de la raison dialectique*, Paris, 1960, p. 17), it must constitute a formidable objection. (That Sartre and Aron, though political opponents, are in full agreement on this point is noteworthy. It shows to what an extent Hegel's concept of history dominates the thought of Marxists and non-Marxists alike.)

Sartre himself, in his *Critique of Dialectical Reason,* gives a kind of Hegelian explanation for his espousal of violence. His point of departure is that "need and scarcity determined the Manicheistic basis of action and morals" in present history, "whose truth is based on scarcity [and] must manifest itself in an antagonistic reciprocity between classes." Aggression is the consequence of need in a world where "there is not enough for all." Under such circumstances, violence is no longer a marginal phenomenon. "Violence and counterviolence are perhaps contingencies, but they are contingent necessities, and the imperative consequence of any attempt to destroy this inhumanity is that in destroying in the adversary the inhumanity of the contraman, I can only destroy in him the humanity of man, and realize in me his inhumanity. Whether I kill, torture, enslave . . . my aim is to suppress his freedom—it is an alien force, *de trop*." His model for a condition in which "each one is one too many . . . Each is *redundant* for the other" is a bus queue, the members of which obviously "take no notice of each other except as a number in a quantitative series." He concludes, "They reciprocally deny any link between each of their inner worlds." From this, it follows that praxis "is the negation of alterity, which is itself a negation"—a highly welcome conclusion, since the negation of a negation is an affirmation.

The flaw in the argument seems to me obvious. There is all the difference in the world between "not taking notice" and "denying," between "denying any link" with somebody and "negating" his otherness; and for a sane person there is still a considerable distance to travel from this theoretical "negation" to killing, torturing, and enslaving.

Most of the above quotations are drawn from R. D. Laing and D. G. Cooper, *Reason and Violence. A Decade of Sartre's Philosophy, 1950-1960*, London, 1964, Part Three. This seems legitimate because Sartre in his foreword says: *"J'ai lu attentivement l'ouvrage que vous avez bien voulu me confier et j'ai eu le grand plaisir d'y trouver un exposé très clair et très fidèle de ma pensée."*

III, TO PAGE 15, NOTE 20

They are indeed a mixed lot. Radical students congregate easily with dropouts, hippies, drug addicts, and psychopaths. The situation is further complicated by the insensitivity of the established powers to the often subtle distinctions between crime and irregularity, distinctions that are of great importance. Sit-ins and occupations of buildings are not the same as arson or armed revolt, and the difference is not just one of degree. (Contrary to the opinion of one member of Harvard's Board of Trustees, the occupation of a university building by students is not the same thing as the invasion of a branch of the First National City Bank by a street mob, for the simple reason that the students trespass upon a property whose use, to be sure, is subject to rules, but to which they belong and which belongs to them as much as to faculty and administration.) Even more alarming is the inclination of faculty as well as administration to treat drug addicts and criminal elements (in City College in New York and in Cornell University) with considerably more leniency than the authentic rebels.

Helmut Schelsky, the German social scientist, described as early as 1961 (in *Der Mensch in der wissenschaftlichen Zivilisation,* Köln und Opladen, 1961) the possibility of a "metaphysical nihilism," by which he meant the radical social and spiritual denial of "the whole process of man's scientific-technical reproduction," that is, the no said to "the rising world of a scientific civilization." To call this attitude "nihilistic" presupposes an acceptance of the modern world as the only possible world. The challenge of the young rebels concerns precisely this point. There is indeed much sense in turning the tables and stating, as Sheldon Wolin and John Schaar have done in *op. cit.:* "The great danger at present is that the established and the respectable . . . seem prepared to follow the most profoundly nihilistic denial possible, which is the denial of the future through denial of their own children, the bearers of the future."

Nathan Glazer, in an article, "Student Power at Berkeley," in *The Public Interest*'s special issue *The Universities,* Fall, 1968, writes: "The student radicals . . . remind me more of the Luddite machine smashers than the Socialist trade unionists who achieved citizenship and power for workers," and he concludes from this impression that Zbigniew Brzezinski (in an article about Columbia in *The New Republic,* June 1, 1968) may have been right in his diagnosis: "Very frequently revolutions are the last spasms of the past, and thus are not really revolutions but counter-revolutions, operating in the name of revolutions." Is not this bias in favor of marching forward at any price rather odd in two authors who are generally considered to be conservatives? And is it not even odder that Glazer should remain unaware of the decisive differences between manufacturing machinery in early nineteenth-century England and the hardware developed in the middle of the twentieth century which has turned out to be destructive even when it appeared to be most beneficial—the discovery of nuclear energy, automation, medicine whose healing powers have led to overpopulation, which in its turn will almost certainly lead to mass starvation, air pollution, et cetera?

IV, TO PAGE 16, NOTE 23

To look for precedents and analogies where there are none, to avoid reporting and reflecting on what is being done and what is being said in terms of the events themselves, under the pretext that we ought to learn the lessons of the past, particularly of the era between the two world wars, has become characteristic of a great many current discussions. Entirely free of this form of escapism is Stephen Spender's splendid and wise report on the student movement, quoted above. He is among the few of his generation to be fully alive to the present *and* to remember his own youth well enough to be aware of the differences in mood, style, thought, and action. ("Today's students are entirely different from the Oxbridge, Harvard, Princeton or Heidelberg students forty years back," p. 165.) But Spender's attitude is shared by all those, in no matter which generation, who are truly concerned with the world's and man's future as distinguished from those who play games with it. (Wolin and Schaar, *op. cit.,* speak of "the revival of a sense of shared destiny" as a bridge between the generations, of "our common fears that scientific weapons may destroy all life, that technology will increasingly disfigure men who live in the city, just as it has already debased the earth and obscured the sky"; that "the 'progress' of industry will destroy the possibility of interesting work; and that 'communica-

tions' will obliterate the last traces of the varied cultures which have
been the inheritance of all but the most benighted societies.") It
seems only natural that this should be true more frequently of
physicists and biologists than of members of the social sciences, even
though the students of the former faculties were much slower to rise
in rebellion than their fellow classmates in the humanities. Thus
Adolf Portmann, the famous Swiss biologist, sees the gap between
the generations as having little if anything to do with a conflict
between Young and Old; it coincides with the rise of nuclear science;
"the resulting world situation is entirely new. . . . [It] cannot be
compared to even the most powerful revolution of the past." (In a
pamphlet entitled *Manipulation des Menschen als Schicksal und
Bedrohung*, Zürich, 1969.) And Nobel Prize winner George Wald, of
Harvard, in his famous speech at M.I.T. on March 4, 1969, rightly
stressed that such teachers understand "the reasons of [their students'] uneasiness even better than they do," and, what is more, that
they "share it," *op. cit.*

V, TO PAGE 17, NOTE 25

The present politicization of the universities, rightly deplored, is
usually blamed on the rebellious students, who are accused of attack-
ing the universities because they constitute the weakest link in the
chain of established power. It is perfectly true that the universities
will not be able to survive if "intellectual detachment and the disinterested search for truth" should come to an end; and, what is
worse, it is unlikely that civilized society of any kind will be able to
survive the disappearance of these curious institutions whose main
social and political function lies precisely in their impartiality and
independence from social pressure and political power. Power and
truth, both perfectly legitimate in their own rights, are essentially
distinct phenomena and their pursuit results in existentially different
ways of life. Zbigniew Brzezinski, in "America in the Technotronic
Age" (*Encounter,* January, 1968), sees this danger but is either re-
signed or at least not unduly alarmed by the prospect. Technotron-
ics, he believes, will usher in a new " 'superculture' " under the
guidance of the new "organization-oriented, application-minded intellectuals." (See especially Noam Chomsky's recent critical analysis
"Objectivity and Liberal Scholarship" in *op. cit.*) Well, it is much
more likely that this new breed of intellectuals, formerly known
as technocrats, will usher in an age of tyranny and utter sterility.
However that may be, the point is that the politicization of the

universities by the students' movement was preceded by the politicization of the universities by the established powers. The facts are too well known to need emphasizing, but it is good to keep in mind that this is not merely a matter of military research. Henry Steele Commager recently denounced "the University as Employment Agency" (*The New Republic*, February 24, 1968). Indeed, "by no stretch of the imagination can it be alleged that Dow Chemical Company, the Marines or the CIA are educational enterprises," or institutions whose goal is a search for truth. And Mayor John Lindsay raised the question of the university's right to call "itself a special institution, divorced from worldly pursuits, while it engages in real-estate speculation and helps plan and evaluate projects for the military in Vietnam" (New York *Times,* "The Week in Review," May 4, 1969). To pretend that the university is "the brain of society" or of the power structure is dangerous, arrogant nonsense —if only because society is not a "body," let alone a brainless one.

In order to avoid misunderstandings: I quite agree with Stephen Spender that it would be folly for the students to wreck the universities (although they are the only ones who could do so effectively for the simple reason that they have numbers, and therefore real power, on their side), since the campuses constitute not only their real, but also their only possible basis. "Without the university, there would be no students" (p. 22). But the universities will remain a basis for the students only so long as they provide the only place in society where power does not have the last word—all perversions and hypocrisies to the contrary notwithstanding. In the present situation, there is a danger that either students or, as in the case of Berkeley, the powers-that-be will run amuck; if this should happen, the young rebels would have simply spun one more thread into what has been aptly called "the pattern of disaster." (Professor Richard A. Falk, of Princeton.)

VI, TO PAGE 19, NOTE 30

Fred M. Hechinger, in an article, "Campus Crisis," in the New York *Times,* "The Week in Review" (May 4, 1969), writes: "Since the demands of the black students especially are usually justified in substance . . . the reaction is generally sympathetic." It seems characteristic of present attitudes in these matters that James Forman's "Manifesto to the White Christian Churches and the Jewish Synagogues in the United States and all other Racist Institutions," though publicly read and distributed, hence certainly "news that's fit to print," remained unpublished until the *New York Review of Books* (July 10, 1969) printed it without the Introduction. Its content, to be sure, is half-illiterate fantasy, and may not be meant seriously. But it is more than a joke, and that the Negro community moodily indulges today in such fantasies is no secret. That the authorities should be frightened is understandable. What can neither be understood nor condoned is their lack of imagination. Is it not obvious that Mr. Forman and his followers, if they find no opposition in the community at large and even are given a little appeasement money, will be forced to try to execute a program which they themselves perhaps never believed in?

VII, TO PAGE 19, NOTE 31

In a letter to the New York *Times* (dated April 9, 1969), Lynd mentions only "nonviolent disruptive actions such as strikes and sit-ins," ignoring for his purposes the tumultuous violent riots of the working class in the twenties, and raises the question why these tactics "accepted for a generation in labor-management relations . . . are rejected when practiced on a campus? . . . when a union organizer is fired from a factory bench, his associates walk off the job until the grievance is settled." It looks as though Lynd has accepted a university image, unfortunately not unfrequent among trustees and administrators, according to which the campus is owned by the board of trustees, which hires the administration to manage their property, which in turn hires the faculty as employees to serve its customers, the students. There is no reality that corresponds to this "image." No matter how sharp the conflicts may become in the academic world, they are not matters of clashing interests and class warfare.

VIII, TO PAGE 19, NOTE 32

Bayard Rustin, the Negro civil-rights leader, has said all that needed to be said on the matter: College officials should "stop capitulating to the stupid demands of Negro students"; it is wrong if one group's "sense of guilt and masochism permits another segment of society to hold guns in the name of justice"; black students were "suffering from the shock of integration" and looking for "an easy way out of their problems"; what Negro students need is "remedial training" so that they "can do mathematics and write a correct sentence," not "soul courses." (Quoted from the *Daily News*, April 28, 1969.) What a reflection on the moral and intellectual state of society that much courage was required to talk common sense in these matters! Even more frightening is the all too likely prospect that, in about five or ten years, this "education" in Swahili (a nineteenth-century kind of no-language spoken by the Arab ivory and slave caravans, a hybrid mixture of a Bantu dialect with an enormous vocabulary of Arab borrowings; see the Encyclopaedia Britannica, 1961), African literature, and other nonexistent subjects will be interpreted as another trap of the white man to prevent Negroes from acquiring an adequate education.

IX, TO PAGE 21, NOTE 36

James Forman's "Manifesto" (adopted by the National Black Economic Development Conference), which I mentioned before and which he presented to the Churches and Synagogues as "only a beginning of the reparations due us as people who have been exploited and degraded, brutalized, killed and persecuted," reads like a classical example of such futile dreams. According to him, "it follows from the laws of revolution that the most oppressed will make the revolution," whose ultimate goal is that "we must assume leadership, total control . . . inside of the United States of everything that exists. The time has passed when we are second in command and the white boy stands on top." In order to achieve this reversal, it will be necessary "to use whatever means necessary, including the use of force and power of the gun to bring down the colonizer." And while he, in the name of the community (which, of course, stands by no means behind him), "declares war," refuses to "share power with whites," and demands that "white people in this country . . . be willing to accept black leadership," he calls at the same time "upon all Christians and Jews to practice patience, tolerance, understanding and nonviolence" during the period it may still take—"whether it happens in a thousand years is of no consequence"—to seize power.

X, TO PAGE 24, NOTE 40

Jürgen Habermas, one of the most thoughtful and intelligent social scientists in Germany, is a good example of the difficulties these Marxists or former Marxists find in parting with any piece of the work of the master. In his recent *Technik und Wissenschaft als 'Ideologie'* (Frankfurt, 1968), he mentions several times that certain "key categories of Marx's theory, namely, class-struggle and ideology, can no longer be applied without ado (*umstandslos*)." A comparison with the essay of Andrei D. Sakharov quoted above shows how much easier it is for those who look on "capitalism" from the perspective of the disastrous Eastern experiments to discard outworn theories and slogans.

XI, TO PAGE 41, NOTE 62

The sanctions of the laws, which, however, are not their essence, are directed against those citizens who—without withholding their support—wish to make an exception for themselves; the thief still expects the government to protect his newly acquired property. It has been noted that in the earliest legal systems there were no sanctions whatsoever. (See Jouvenel, *op. cit.*, p. 276.) The lawbreaker's punishment was banishment or outlawry; by breaking the law, the criminal had put himself outside the community constituted by it.

Passerin d'Entrèves (*op. cit.*, pp. 128 ff.), taking into account "the complexity of law, even of State law," has pointed out that "there are indeed laws which are 'directives' rather than 'imperatives', which are 'accepted' rather than 'imposed', and whose 'sanctions' do not necessarily consist in the possible use of force on the part of a 'sovereign'." Such laws, he has likened to "the rules of a game, or those of my club, or to those of the Church." I conform "because for me, unlike others of my fellow citizens, these rules are 'valid' rules."

I think Passerin d'Entrèves's comparison of the law with the "valid rules of the game" can be driven further. For the point of these rules is not that I submit to them voluntarily or recognize theoretically their validity, but that in practice I cannot enter the game unless I conform; my motive for acceptance is my wish to play, and since men exist only in the plural, my wish to play is identical with my wish to live. Every man is born into a community with pre-existing laws which he "obeys" first of all because there is no other way for him to enter the great game of the world. I may wish to change the rules of the game, as the revolutionary does, or to make an exception for myself, as the criminal does; but to deny them on principle means no mere "disobedience," but the refusal to enter the human community. The common dilemma—either the law is absolutely valid and therefore needs for its legitimacy an immortal, divine legislator, or the law is simply a command with nothing behind it but the state's monopoly of violence—is a delusion. All laws are " 'directives' rather than 'imperatives.' " They direct human intercourse as the rules direct the game. And the ultimate guarantee of their validity is contained in the old Roman maxim *Pacta sunt servanda.*

XII, TO PAGE 50, NOTE 72

There is some controversy on the purpose of de Gaulle's visit. The evidence of the events themselves seems to suggest that the price he had to pay for the army's support was public rehabilitation of his enemies—amnesty for General Salan, return of Bidault, return also of Colonel Lacheroy, sometimes called the "torturer in Algeria." Not much seems to be known about the negotiations. One is tempted to think that the recent rehabilitation of Pétain, again glorified as the "victor of Verdun," and, more importantly, de Gaulle's incredible, blatantly lying statement immediately after his return, blaming the Communist party for what the French now call *les événements,* were part of the bargain. God knows, the only reproach the government could have addressed to the Communist party and the trade unions was that they lacked the power to prevent *les événements.*

XIII, TO PAGE 54, NOTE 75

It would be interesting to know if, and to what an extent, the alarming rate of unsolved crimes is matched not only by the well-known spectacular rise in criminal offenses but also by a definite increase in police brutality. The recently published *Uniform Crime Report for the United States,* by J. Edgar Hoover (Federal Bureau of Investigation, United States Department of Justice, 1967), gives no indication how many crimes are actually solved—as distinguished from "cleared by arrest"—but does mention in the Summary that police solutions of serious crimes declined in 1967 by 8%. Only 21.7 (or 21.9)% of all crimes are "cleared by arrest," and of these only 75% could be turned over to the courts, where only about 60% of the indicted were found guilty! Hence, the odds in favor of the criminal are so high that the constant rise in criminal offenses seems only natural. Whatever the causes for the spectacular decline of police efficiency, the decline of police power is evident, and with it the likelihood of brutality increases. Students and other demonstrators are like sitting ducks for police who have become used to hardly ever catching a criminal.

A comparison of the situation with that of other countries is difficult because of the different statistical methods employed. Still, it appears that, though the rise of undetected crime seems to be a fairly general problem, it has nowhere reached such alarming proportions as in America. In Paris, for instance, the rate of solved crimes declined from 62% in 1967 to 56% in 1968, in Germany from 73.4% in 1954 to 52.2% in 1967, and in Sweden 41% of crimes were solved in 1967. (See "Deutsche Polizei," in *Der Spiegel*, April 7, 1967.)

XIV, TO PAGE 55, NOTE 76

Solzhenitsyn shows in concrete detail how attempts at a rational economic development were wrecked by Stalin's methods, and one hopes this book will put to rest the myth that terror and the enormous losses in human lives were the price that had to be paid for rapid industrialization of the country. Rapid progress was made after Stalin's death, and what is striking in Russia today is that the country is still backward in comparison not only with the West but also with most of the satellite countries. In Russia there seems not much illusion left on this point, if there ever was any. The younger generation, especially the veterans of the Second World War, knows very well that only a miracle saved Russia from defeat in 1941, and that this miracle was the brutal fact that the enemy turned out to be even worse than the native ruler. What then turned the scales was that police terror abated under the pressure of the national emergency; the people, left to themselves, could again gather together and generate enough power to defeat the foreign invader. When they returned from prisoner-of-war camps or from occupation duty they were promptly sent for long years to labor and concentration camps in order to break them of the habits of freedom. It is precisely this generation, which tasted freedom during the war and terror afterward, that is challenging the tyranny of the present regime.

XV, TO PAGE 66, NOTE 86

No one in his right senses can believe—as certain German student groups recently theorized—that only when the government has been forced "to practice violence openly" will the rebels be able "to fight against this shit society (*Scheissgesellschaft*) with adequate means and destroy it." (Quoted in *Der Spiegel*, February 10, 1969, p. 30.) This linguistically (though hardly intellectually) vulgarized new version of the old Communist nonsense of the thirties, that the victory of fascism was all to the good for those who were against it, is either sheer play-acting, the "revolutionary" variant of hypocrisy, or testifies to the political idiocy of "believers." Except that forty years ago it was Stalin's deliberate pro-Hitler policy and not just stupid theorizing that stood behind it.

To be sure, there is no reason for being particularly surprised that German students are more given to theorizing and less gifted in political action and judgment than their colleagues in other, politically more fortunate, countries; nor that "the isolation of intelligent and vital minds . . . in Germany" is more pronounced, the polarization more desperate, than elsewhere, and their impact upon the political climate of their own country, except for backlash phenomena, almost nil. I also would agree with Spender (see "The Berlin Youth Model," in *op. cit.*) about the role played in this situation by the still-recent past, so that the students "are resented, not just on account of their violence, but because they are reminders . . . they also have the look of ghosts risen from hastily covered graves." And yet, when all this has been said and duly taken into account, there remains the strange and disquieting fact that none of the new leftist groups in Germany, whose vociferous opposition to nationalist or imperialist policies of other countries has been notoriously extremist, has concerned itself seriously with the recognition of the Oder-Neisse Line, which, after all, is the crucial issue of German foreign policy and the touchstone of German nationalism since the defeat of the Hitler regime.

XVI, TO PAGE 73, NOTE 99

Daniel Bell is cautiously hopeful because he is aware that scientific and technical work depend on "theoretical knowledge [that] is sought, tested, and codified in a disinterested way" (*op. cit*). Perhaps this optimism can be justified so long as the scientists and technologists remain uninterested in power and are concerned with no more than social prestige, that is, so long as they neither rule nor govern. Noam Chomsky's pessimism, "neither history nor psychology nor sociology gives us any particular reason to look forward with hope to the rule of the new mandarins," may be excessive; there are as yet no historical precedents, and the scientists and intellectuals who, with such deplorable regularity, have been found willing to serve every government that happened to be in power, have been no "meritocrats" but, rather, social climbers. But Chomsky is entirely right in raising the question: "Quite generally, what grounds are there for supposing that those whose claim to power is based on knowledge and technique will be more benign in their

100

exercise of power than those whose claim is based on wealth or aristocratic origin?" (*Op. cit.*, p. 27.) And there is every reason to raise the complementary question: What grounds are there for supposing that the resentment against a meritocracy, whose rule is exclusively based on "natural" gifts, that is, on brain power, will be no more dangerous, no more violent than the resentment of earlier oppressed groups who at least had the consolation that their condition was caused by no "fault" of their own? Is it not plausible to assume that this resentment will harbor all the murderous traits of a racial antagonism, as distinguished from mere class conflicts, inasmuch as it too will concern natural data which cannot be changed, hence a condition from which one could liberate oneself only by extermination of those who happen to have a higher I.Q.? And since in such a constellation the numerical power of the disadvantaged will be overwhelming and social mobility almost nil, is it not likely that the danger of demagogues, of popular leaders, will be so great that the meritocracy will be forced into tyrannies and despotism?

XVII, TO PAGE 77, NOTE 106

Stewart Alsop, in a perceptive column, "The Wallace Man," in *Newsweek*, October 21, 1968, makes the point: "It may be illiberal of the Wallace man not to want to send his children to bad schools in the name of integration, but it is not at all unnatural. And it is not unnatural either for him to worry about the 'molestation' of his wife, or about losing his equity in his house, which is all he has!" He also quotes the most effective statement of George Wallace's demagoguery: "There are 535 members of Congress and a lot of these liberals have children, too. You know how many send their kids to the public schools in Washington? Six."

Another prime example of ill-designed integration policies was recently published by Neil Maxwell in *The Wall Street Journal* (August 8, 1968). The federal government promotes school integration in the South by cutting off federal funds in cases of flagrant noncompliance. In one such instance, $200,000 of annual aid was withheld. "Of the total, $175,000 went directly to Negro schools. . . . Whites promptly raised taxes to replace the other $25,000." In short, what is supposed to help Negro education actually has a "crushing impact" on their existing school system and no impact at all on white schools.

In the murky climate of ideological talk and doubletalk of Western student debate, these issues seldom have a chance of being clarified; indeed, "this community, verbally so radical, has always sought and found an escape," in the words of Günter Grass. It is also true that this is especially noticeable and infuriating in German students and other members of the New Left. "They don't know anything, but they know it all," as a young historian in Prague, according to Grass, summed it up. Hans Magnus Enzensberger gives voice to the general German attitude; the Czechs suffer from "an extremely limited horizon. Their political substance is meager." (See Günter Grass, *op. cit.*, pp. 138-142.) In contrast to this mixture of stupidity and impertinence, the atmosphere among the eastern rebels is refreshing, although one shudders to think of the exorbitant price that has been paid for it. Jan Kavan, a Czech student leader, writes: "I have often been told by my friends in western Europe that we are only fighting for bourgeois-democratic freedoms. But somehow I cannot seem to distinguish between capitalist freedoms and socialist freedoms. What I recognize are basic human freedoms." (*Ramparts*, September 1968.) It is safe to assume that he would have a similar difficulty with the distinction between "progressive and repressive violence." However, it would be wrong to conclude, as is so frequently done, that people in the western countries have no legitimate complaints precisely in the matter of freedom. To be sure, it is only natural "that the attitude of the Czech to the western students is largely coloured by envy" (quoted from a student paper by Spender, *op. cit.*, p. 72), but it is also true that they lack certain, less brutal and yet very decisive experiences in political frustration.

Index

Algeria, 14, 53, 98
Alsop. Stewart, 101
American Political Science Review, The, 72n
Aron, Raymond, 49n, 89-90

Bakunin, Mikhail, 89
Barion, Jacob, 12n
Barnes, Peter, 29n
Beaufre, André, 5n
Bell, Daniel, 73, 100
Bergson, Henri, 12, 70, 73, 74
Berlin, Isaiah, 27n
Bidault, Georges, 98
Bodin, Jean, 38
Böll, Heinrich, 46n
Borkenau, Franz, 47n-48n
Brzezinski, Zbigniew, 92, 93

Calder, Nigel, 3n, 5n, 10n, 62n
Castro, Fidel, 21n
Catherine the Great, 74
China, 21n
Chomsky, Noam, 7, 7n, 14n, 23n, 48n, 64n, 93, 100-101
Cicero, 43n
City College of New York, 18, 91
Clausewitz, Karl von, 8-9, 10, 36, 37n
Columbia University, 79, 92
Commager, Henry Steele, 17, 54, 85n, 94

Commonweal, 23
Commentary, 28n
Cooper, D. G., 91
Cornell University, 18, 91
Cromer, Lord, 54
Cuba, 21n
Czechoslovakia, 24, 53, 83, 102

Dedijer, Vladimir, 10n
de Gaulle, Charles, 50, 98
Deming, Barbara, 71n
DesJardins, Gregory, 40n
Dreyfus, Alfred, 71
Dutschke, Rudi, 79n

Ehmann, Christoph, 25n
Encounter, 93
Encyclopaedia Britannica, 96
Encyclopedia of the Social Sciences, 8
Engels, Friedrich, 4, 9, 12n, 20, 22n, 89
England, 38, 53, 54, 92
d'Entrèves, Alexander Passerin, 37-38, 43, 97
Enzensberger, Hans Magnus, 102
Europe, 6, 15, 53, 69, 102

Falk, Richard A., 94
Fanon, Frantz, 12, 14, 20, 21n, 65, 67, 69, 71, 75, 90

Feuerbach, Ludwig, 89
Finer, S. E., 72n, 82n
Fogelson, Robert M., 76n
Fontenelle, Bernard le Bovier de, 25
Forman, James, 77, 94-95, 96
France, 15, 38, 53, 79, 98, 99
Fulbright, William, 16-17

Gandhi, Mahatma, 53
Gans, Herbert J., 83n
Germany, 15, 18, 23n, 24-25, 42, 50, 53, 79, 96, 99, 100, 102
Glazer, Nathan, 91-92
Goodwin, Richard N., 7
Grass, Günter, 79n-80n, 83n, 102
Gray, J. Glenn, 67n
Greece, 50n
Guevara, Che, 21n

Habermas, Jürgen, 96
Halevy, Daniel, 70n
Harbold, William H., 26n
Harvard University, 91, 92, 93
Hechinger, Fred M., 79n, 94
Hegel, Georg Friedrich, 12-13, 26, 27, 28, 56, 90
Heidelberg University, 92
Herzen, Alexander, 27
Hitler, Adolf, 42, 53, 100
Hobbes, Thomas, 5, 38, 68
Ho Chi Minh, 21
Hoover, J. Edgar, 98
Hull University, 89

India, 53

Japan, 15, 53
Jouvenel, Bertrand de, 36, 37, 38, 41, 74n, 97

Kant, Immanuel, 27
Kavan, Jan, 102
Klineberg, Otto, 62n
Kohout, Pavel, 83

Lacheroy, Colonel, 98
Laing, R. D., 91
Lenin, Nikolai, 3, 12, 22, 24
Lessing, Gotthold Ephraim, 25
Lettvin, Jerome, 16n, 17
Liberation, 71n
Lindsay, John V., 94
Litten, Jens, 81
Lorenz, Konrad, 61n, 69
Lübke, Karl-Heinz, 45n
Lynd, Staughton, 19, 95

Madison, James, 41
Mao Tse-tung, 11, 21n, 38
Massu, Jacques, 50
Marx, Karl, 11, 12, 13, 15n, 20, 21, 22, 24, 25, 26, 27, 36, 56, 72, 74, 89, 96
Massachusetts Institute of Technology, 16, 17, 93
Maxwell, Neil, 101
McCarthy, Eugene, 85-86
McIver, R. M., 37n
Melville, Herman, 64
Mill, John Stuart, 39, 40
Mills, C. Wright, 35
Mitscherlich, Alexander, 61n
Montesquieu, Charles L. de, 41

Napoleon Bonaparte, 51, 79
National Black Economic Development Conference, 96
National Guard, 29, 54
National Liberation Front, 48

104

Nechaev, Sergey Kravinsky, 89
New Left, 11, 13-14, 23, 89, 102
New Republic, The, 54n, 92, 94
New York *Daily News*, 95
New York Review of Books, 16n, 89, 95
New York *Times*, 19n, 46n, 79n, 94, 95
New York *Times Magazine*, 16n
New Yorker, The, 7n, 18n
Newsweek, 29n, 101
Nietzsche, Friedrich, 44, 73, 74
Nisbet, Robert A., 28n
Nixon, Richard, 86

O'Brien, Conor Cruise, 79
O'Brien, William, 79

Parekh, B. C., 89
Pareto, Vilfredo, 65, 71, 72, 81, 82
Pascal, Blaise, 25
Péguy, Charles, 23
Pétain, Henri Philippe, 98
Plato, 44
Portmann, Adolf, 60, 93
Princeton University, 92, 94
Proudhon, P.-J., 7, 26
Public Interest, The, 73n, 91

Ramparts, 102
Rand Corporation, 5n
Renan, Joseph Ernest, 8
Review of Politics, 26n
Robespierre, Maximilien, 65
Rockefeller, Nelson A., 86
Rousseau, Jean Jacques, 37n
Russia, 52-53, 55-56, 99
Rustin, Bayard, 95

Sakharov, Andrei D., 9-10, 46n, 83n, 97
Salan, Raoul, 98
Sartre, Jean-Paul, 12, 13, 20, 21, 36, 89, 90-91
Schaar, John, 16n, 29n, 45n, 91, 92-93
Schapiro, Leonard, 89
Schelling, Thomas C., 7n
Schelsky, Helmut, 91
Science, 60n
Selden, John, 84
Silver, Allan A., 77n
Solzhenitsyn, Aleksandr I., 55, 99
Sorel, Georges, 12, 20, 35, 65, 69, 70, 71, 72, 81
Spain, 51
Spender, Stephen, 17n, 18, 21n, 23n, 29n, 42, 49n-50n, 66n, 92, 94, 100, 102
Spengler, Oswald, 69
Spiegel, Der, 14n-15n, 23n, 25n, 79n, 80n, 99
Stalin, Joseph, 53, 55, 99, 100
Steinfels, Peter, 23
Strausz-Hupé, Robert, 36-37
Students for a Democratic Society (SDS), 66n
Sweden, 99

Theatre of Ideas, 79
Thring, M. W., 10n
Tinbergen, Nikolas, 60n, 62n
Tito, Josip Broz, 21n
Trotsky, Leon, 35n

United States, 5-6, 15, 18, 19, 24, 29n, 48, 80, 85-86, 98-99

University of California (Berkeley), 16n, 28-29, 45n, 54, 91, 92, 94
Valéry, Paul Ambroise, 86-87
Venturi, Franco, 27n
Verdun, 98
Vietnam, 14, 48, 51, 54, 64n, 94
Voltaire, François, 36
von Holst, Erich, 60n, 62n

Wald, George, 18, 93
Wallace, George, 101

Wallace, John M., 40n
Wall Street Journal, The, 101
Weber, Max, 35, 36, 37n
Wheeler, Harvey, 3n, 9n, 10n
Wilson, Edmund, 30n
Wilson, James, 6
Wolin, Sheldon, 16n, 29n, 45n, 91, 92-93

Xenophon, 50n

Yugoslavia, 21n